Dulcinéia Cristina Chiapetti
Mafalda Nesi Francischett

Teaching Topological Relationships to children in the Early Years

Dulcinéia Cristina Chiapetti
Mafalda Nesi Francischett

Teaching Topological Relationships to children in the Early Years

Orientation and Localisation in Primary School

ScienciaScripts

Imprint

Any brand names and product names mentioned in this book are subject to trademark, brand or patent protection and are trademarks or registered trademarks of their respective holders. The use of brand names, product names, common names, trade names, product descriptions etc. even without a particular marking in this work is in no way to be construed to mean that such names may be regarded as unrestricted in respect of trademark and brand protection legislation and could thus be used by anyone.

Cover image: www.ingimage.com

This book is a translation from the original published under ISBN 978-613-9-66922-6.

Publisher:
Sciencia Scripts
is a trademark of
Dodo Books Indian Ocean Ltd. and OmniScriptum S.R.L publishing group

120 High Road, East Finchley, London, N2 9ED, United Kingdom
Str. Armeneasca 28/1, office 1, Chisinau MD-2012, Republic of Moldova, Europe
Printed at: see last page
ISBN: 978-620-8-11176-2

ACKNOWLEDGEMENTS

To the University of Western Paraná - UNIOESTE, Francisco Beltrão Campus, for their contribution to my education. To my supervisor Prof Mafalda Nesi Francischett for always helping me in times of doubt. To my work colleagues who have always encouraged and supported me at this stage of my professional training, especially Professor Junice Forner (in memorian) who was unable to see this work through to completion. To my family, my friends and my fellow students.

SUMMARY

This dissertation on "Teaching topological relationships to children in the early years of **primary school" focuses on** the difficulties children have in the learning process, especially in forming concepts involving laterality, cardinal directions and other guiding principles. It is justified by the fact that many children have difficulties understanding topological relationships and relating lateral directions to cardinal directions (right/east, left/west, front/north and back/south). Also because of the difficulty of teaching geography in early childhood education and, consequently, the lack of importance given to geographical content. The text addresses the Activity Theory proposed by Leontiev, because it supports the foundations of the relationship between those who teach and those who learn. It explains the process of constructing concepts and understanding certain activities carried out in the school environment. Another basis is Vygotsky because he emphasises action and learning as a process that occurs from the outside in and is full of historicity. Also because of the context in which the teacher mediates the process. Another theorist is Luria, who advocates activities in a concrete context, to make it possible to move on to abstract categories. This methodological case study research is based on Yin (2001). Data collection was based on interviews with 114 children aged between five (5) and six (6); five (5) teachers; one (1) coordinator. It was carried out in four (4) municipal schools in Itapejara **D'Oeste/PR. The interviews were carried out individually and** 39 activities **were** worked on, four (4) individually and thirty-five (35) collectively. It provided a space for discovering teaching methodologies through which teachers can teach important geographical concepts. It identifies the importance of considering the individuality of the child in order to investigate their difficulties. The didactic sequences give children the opportunity to participate in the construction of knowledge, and show the importance of pedagogical work related to the geographical concepts of: laterality; cardinal directions, as well as working on cartographic representations. If children have difficulty forming these concepts, it's because the process is still underway. The child needs to have the knowledge associated with something that is part of their everyday life, especially with regard to cardinal directions and spatial references. When children don't understand the relationship between the concepts north/front, south/back, east/right and west/left, it becomes difficult for them to orientate or locate themselves in space and even to understand what these concepts are for.
Keywords: Geography teaching. Laterality. Cardinal directions.

INTRODUCTION

For children to develop their geographical thinking in the early years of primary school, they need to learn some basic concepts in this area of knowledge. Thus, topological relationships characterised as knowledge of cardinal directions and laterality make a significant contribution to understanding, analysing and interpreting geographical space. This research is based on the principle that actions that relate classroom practices to the teaching of topological relationships help children to develop and formulate important concepts in Geography.

The different ways of teaching topological relationships help children to relate objects, people and places as a reference. These contents help in the process of articulating and formulating the concepts necessary for appropriating information and developing reasoning, as they require spatial contextualisation.

Activities illustrated with images and sounds are common in the early years and are important for comprehension. For example, the map legend, in which the colour variable, often used in drawings, helps to convey correlated information. Pre-school children begin to acquire the notions of proportion, reduction and inclusion, which are essential for raising their thinking from the abstract level to understanding concepts at the concrete level. For example, scale and legend, in order to analyse the differences and similarities between what is actually experienced and what is represented on maps.

The theoretical basis for this dissertation is mainly Vygotsky[1] , Leontiev and Luria. It uses Activity Theory to explain the process of constructing concepts, in order to understand certain activities carried out in the school environment with children.

This methodological case study, based on Yin (2001), presents actions that prioritise and promote experimentation in children. Data collection was based on interviews with one hundred and fourteen (114) children[2] , five (5) teachers and one (1) coordinator in four (4) municipal schools in Itapejara D'Oeste/PR. Thirty-nine (39) interviews were carried out, four (4) individually and thirty-five (35) collectively. Teachers and coordinators took part in presenting and evaluating the actions carried out in everyday school life.

The interviews began in the first half of 2016 and were completed at the end of the second half of 2017. At first, the research involved twenty-six (26) pre-school children and twenty-five (25) first graders enrolled at the Nereu Ramos Municipal School. Subsequently, due to the need to identify the formation of concepts, they were carried out in the other three (3) schools, involving a further

---

[1] This author is written about in different ways throughout the text because his works have been translated into many different languages.
[2] The names of the children who appear in this research are fictitious.

sixty-three (63) children.

The pedagogical actions were carried out with children aged between five (5) and six (6) years old, who are in the transition phase between pre-school and primary school, and promote the formation of concepts involving laterality as a guiding principle for cardinal directions, which is important for reading cartographic representations in Geography.

The dissertation is presented in three chapters: the first deals with the relationship between teaching, learning and knowledge, with Activity Theory and mediation in the process as its guiding element. The second presents laterality in spatial organisation and the pedagogical actions carried out with pre-school children (2016) and first graders (first half of 2017) at the Nereu Ramos Municipal School. The third chapter discusses the formation of concepts through interviews conducted in the second half of 2017 with children from the Irmão Josafat Kmita, Prof. Pedro Viriato Parigot de Souza and Valentim Biazussi schools.

CHAPTER 1

THE RELATIONSHIP BETWEEN TEACHING, LEARNING AND KNOWLEDGE

1.1 Concept formation at school

Scientific concepts occur differently from spontaneous concepts and require a greater degree of abstraction. For this to happen, children at school need a mediator (teacher). Vygotsky (2009) developed his theory with school environments in mind. His principle is to incorporate social factors as essential to the formation of concepts.

Concepts are seen by Vygotsky (1984) as something subordinate to certain schemas and have, in activity, a means of associating thought with representation. When forming a concept, the child needs to know how to name and signify in order to develop more complex thoughts. This is why concept formation is linked to problem solving. Ivic and Coelho (2010) also focus on Vygotsky and present the development of scientific concepts at school age as one of the school's functions, by placing children in front of a teaching system that promotes the development of scientific concepts.

Pedagogical action is an important indicator of teaching work. This is why it is the guiding question of this dissertation. It is understood that the actions worked on by the teacher help in the formation of concepts and this reflects on the learning process and the acquisition of knowledge.

In this context, it is proposed to adhere to the Activity Theory, developed by Leontiev (1978), which emphasises that development occurs through the need and satisfaction of the child's relationship with the environment in which they are inserted (school). This satisfaction promotes the development of psychic functions and stems from the appropriation of knowledge, which transforms external activity (someone who teaches) into internal activity (someone who learns) mediated by the relationships between the subjects (child/teacher, child/children). He says that you have to have a reason to learn. This is what drives the child to carry out an action through learning. The child needs to know why they are carrying out a particular pedagogical action and where they want to go with it.

Through mediation, it is possible to understand the child and enhance the internalisation of concepts, their development and, consequently, learning. Activity Theory is an essential process for analysing this research, because it is through pedagogical action that the interaction between content, the social context and, in this case, the learning of geographical concepts takes place.

Activity Theory emerged from the work of Vygotsky, Luria and Leontiev and conceives action as the result of the interaction between the objective structure of human activity and the subjective structure of consciousness. Activity is an action present in human beings and through it consciousness is formed, which can be both humanising and alienating (VYGOTSKY, 2000).

In Leontiev's Activity Theory, several concepts were developed by Vygotsky, including the historical construction of the relationship between man and the world. Leontiev (1978) understands mediation as an element of fundamental importance. For him, what differentiates human beings from animals is their ability to plan their actions in order to achieve a goal. Activities represent the way human beings relate to the world and happen intentionally, as they are planned by someone responsible for mediating this action.

> Activity is a molecular unity... it is the unity of life mediated by the psychological reflex, whose real function is to orientate the subject in the objective world. In other words, activity is not a reaction or a set of reactions, but a system that has a structure, its internal transitions and transformations, its development. (LEONTIEV, 1978, p. 66-67).

Material and intellectual culture promotes the appropriation of the real world and the conceptual world. In this sense, activity includes motor and mental actions that favour awareness of processes involving the appropriation of problems to be solved. When carrying out a practical or cognitive activity, the child appropriates instruments and signs. Whether individually or collectively, this produces meaning (internalisation) and, consequently, learning (GUERRERO, 2007).

For Leontiev (1978), the pedagogical actions carried out at school help in the appropriation and reproduction of social actions, as they are processes that are historically and socially constructed by people with less experience. These activities favour psychological development and contribute to the teaching and learning process when used appropriately.

Activity Theory underpins the relationship between those who teach and those who learn, in the role of someone mediating this process. For Leontiev (1978), there are two main characteristics in Activity Theory, one is objectivity and the other is necessity. These two characteristics lead individuals to act, in other words, the activity needs to trigger an action.

> [...] Human activity always has a motive. When this activity becomes composed of smaller units, the actions, this means that each of the individual actions that make up the collective activity no longer has a direct relationship with the motive of the activity and begins to maintain an indirect, mediated relationship with that motive. (DUARTE, 2002, p. 285).

While carrying out the activity, the child develops a relationship between meaning and sense, which are the reasons, or motives, that make them carry out the activity. This action is the psychic reflex presented by the mediated structure of consciousness, which for the teacher has an exchange value between what has been learnt and what has been taught (DUARTE, 2002).

The child's development is linked to their place in social relations, and what determines their psychic development is their evolution. The child necessarily needs to develop action and operation on the basis of function. These are the foundations of the development of individual consciousness (VIGOTSKII; LEONTIEV; LÚRIA, 2010).

Through activity, children develop their perception, which is initially simpler and becomes more complex as they incorporate certain skills throughout their development.

During pedagogical activities, it is essential that the teacher considers the relationship between learning and development (VIGOTSKII; LEONTIEV; LÚRIA, 2010).

Through activity, children develop. However, this process of knowledge is long and difficult. When carrying out an action, the child needs to plan it. At first, activities need to refer to the concrete context. Only later do they move on to abstract categories (VIGOTSKII; LEONTIEV; LÚRIA, 2010).

Vygotskyi, Leontiev and Luria (2010) proposed this theory to develop socially organised experiences, which determine the very structure of conscious human activity.

> Since all activity is initially fixed on graphic and practical operations, we believe that the development of conceptual, taxonomic thinking is linked to the theoretical operations that a child learns to perform at school (VIGOTSKII; LEONTIEV; LÚRIA, 2010, p. 48).

The processes of abstraction and generalisation vary in children, depending on their stage of socio-economic and cultural development. These processes are the result of the cultural environment in which the child is inserted (VIGOTSKII; LEONTIEV; LÚRIA, 2010).

Pre-school is a period in a child's school life when the human reality that surrounds them begins to define itself. In all practical activities, especially games, the child goes beyond the limits of manipulating objects and enters a wider universe, thus assimilating human objects and reproducing them through actions. In this way, children recognise their dependence on the people around them (VIGOTSKII; LEONTIEV; LÚRIA, 2010).

> They have to take into account the demands made on their behaviour by the people around them, because this really determines their personal, intimate relationships with these people. Not only do her successes and failures depend on these relationships, but her joys and sorrows are also involved with these relationships and have the force of motivation. (VIGOTSKII; LEONTIEV; LÚRIA, 2010, p. 60).

Activity as a whole is not built up mechanically or through separate actions. However, over time, the child's knowledge increases along with their abilities. This change gives meaning to the activity and causes a transition from one stage of development to another. In pre-school, the child connects facts corresponding to their potential and this determines changes in their perception of the world (VIGOTSKII; LEONTIEV; LÚRIA, 2010).

> By activity, we mean processes that are psychologically characterised by what the process, as a whole, is directed towards (its object), always coinciding with the objective that stimulates the subject to carry out this activity, in other words, the motive (VIGOTSKII; LEONTIEV; LÚRIA, 2010, p. 68).

There is a particular relationship between activity and action, because the motive of the activity becomes the object (the target) of the action and the result is an action transformed into activity. The knowledge that children possess is their ability to interpret phenomena using their own activity as an instrument (VIGOTSKII; LEONTIEV; LÚRIA, 2010).

The development of the child's consciousness is expressed in the motivation he or she shows in carrying out an activity. We can thus say that, "[...] old motives lose their stimulating force, and new ones are born, leading to a reinterpretation of their previous actions [...]" (VIGOTSKII; LEONTIEV; LÚRIA, 2010, p. 82).

In the same way, the symbol acquires a functional meaning that is reflected in the content presented to the child. The fact that the child develops the skills to answer questions does not mean that he or she has understood the process (VIGOTSKII; LEONTIEV; LÚRIA, 2010).

Abstraction, therefore, is a mental transition from the general to the particular and is extremely important for school learning. The process of abstraction can also be a process of generalisation, as it presents links and relationships between children and the actions presented to them. For children to build concepts, there are essential and non-essential attributes in the activities they are offered (GUERRERO, 2007).

> The concept would be the common abstraction, the mental representation we make of a group of objects. The intuitive (or direct) method, which goes from the particular to the general, provides the basis on which empirical generalisation will take place. The deductive method, which is independent of direct observations and is the result of deductions based on relationships established by subjects with objects and phenomena, is the basis for theoretical generalisations.
> Therefore, empirical generalisation presupposes carrying out analyses of the initial construction conditions of a system of objects through their transformation, allowing individuals, after solving a series of concrete and practical problems, to appropriate knowledge. Theoretical generalisation, on the other hand, consists of valuing the common and externally similar properties of a variety of objects when comparing them. The process of acquiring knowledge in the latter goes beyond empirical experience and requires learning activities with appropriate content and structures (GUERRERO, 2007, p. 121).

Every activity needs a problem to trigger an action, so solving this problem transforms the child's action (GUERRERO, 2007).

The activity indicated by the teacher can also alienate the child. Alienation is also a historically and socially constructed process between subjects. For Vygotsky (2009), the relationship between the spontaneous concept and the scientific concept provides the basis for structuring the teaching and learning process.

There is a historical relationship between the subject and their objective in carrying out an activity in a reciprocal way, permanently evolving until they reach a final result. There is a need for exchange in order to express, through communication (language), this interaction between the subject and internalised knowledge through psychic functions. Speech enables the child to interact with the environment. This ends up internalising the knowledge obtained by

This is reflected in the practical activities carried out in the classroom under the guidance of the teacher (LEONTIEV, 1978).

The construction of knowledge is a shared action which, according to Vygotsky (2000), requires the co-operation of both the teacher and the child because it is based on the development of the child's individual, mental capacities and develops throughout life, through contact with information and interaction with the socio-cultural environment. Unlike animals, human beings have capacities that make it possible to carry out activities, which can be passed on to the group to which they belong and can be perfected so that they can be transformed into new activities.

An activity can be carried out through various actions and be based on different motives with different personal meanings for each actor (RODRIGUES, 2016).

In this sense, "[...] the action-operation dynamic is characteristic of **human development**". (RODRIGUES, 2016, p. 03). In other words, children need to know why they are doing what they are doing and where they can get to. Otherwise, what the teacher does is just a simple task. **"This is why** one of the teacher's apprenticeships is the construction of teaching proposals that can be realised through **educational** activities." (GUERRERO, 2007, p. 125).

The use of activities, in the school context, goes through a process that involves the world of work and education, so it needs continuity in studies to be able to represent its use. (RODRIGUES, 2016). Therefore, "[...] thinking and doing are not located at opposite poles". (RODRIGUES, 2016, p. 03).

The essence of learning lies in the needs and motivations of the learner. Teaching and learning go through paths where aptitudes, motivations, pleasure and willingness to learn must be present (MORETTI; ASBAHR; RIGON, 2011).

Work, as well as the techniques developed over time, defined the human species, which gradually also developed a complex system of language that articulates signs and meanings, structuring the communication process. Gradually, the coexistence of civilisations led to the establishment of rituals, laws, religions, political and economic norms that were standard in each society (MORETTI; ASBAHR; RIGON, 2011).

> Our starting point for explaining the process of humanisation is cultural-historical theory, whose epistemological origins lie in dialectical materialism, based on the works of Marx. This author considers that the human being is the result of the intertwining of the individual aspect, in the biological sense, and the social aspect, in the cultural sense. In other words, by appropriating culture and everything that the human species has developed [...] man becomes human (MORETTI; ASBAHR; RIGON, 2011, p. 478).

One of the presuppositions of cultural-historical theory is that work is a human activity that plays a central role in the development of human beings, humanising them and making it possible for culture to spread. Man creates conditions and needs to maintain his biological and cultural existence. These intentional attitudes establish certain principles that guide their actions, becoming a cultural-historical necessity that fulfils their needs (MORETTI; ASBAHR; RIGON, 2011).

For Sánches Vásquez (1977), what makes man human is the fact that he invents or creates needs and the conditions for satisfying these needs. This action is part of consciousness. For this reason, human activity is intentional and has the capacity for planning of actions. With this, "[...] **man, through work, begins to control his behaviour [...]"** (MORETTI, ASBAHR; RIGON, 2011, p. 479).

However, **this "[...] movement is not individual, but fundamentally collective and responsible for the constitution of culture". (MORETTI, ASBAHR;** RIGON, 2011, p. 479). When living in society, man starts to develop certain individual activities. In this relationship, individuality and genericity become singular agents in society, resulting in historical and cultural production (MORETTI, ASBAHR; RIGON, 2011).

> In the relationship between consciousness and activity, consciousness is the specifically human form of the psychic reflection of reality, in other words, it is the expression of the individual's relationship with the social, cultural and historical world and opens up to man a picture of the world in which he himself is inserted. Consciousness thus refers to the human possibility of understanding the social and individual world as analysable (MORETTI, ASBAHR; RIGON, 2011, p. 479).

As such, language, like other actions, plays a fundamental role in social relations, because men are able to share representations, concepts and techniques with other generations. Activity and consciousness are categories of cultural-historical theory. But it is sensory practical activity that develops individuals (MORETTI; ASBAHR; RIGON, 2011). "In teaching, working with representations presupposes relating them both to the concept and to the connotative aspects that may exist implicitly." (FRANCISCHETT, 2014, p. 844).

According to Vygotsky (2009), pre-school children develop pseudo-concepts that are like bridges linking the concept to thought. For this reason, the grouping of ideas needs to be centred in the visible field and directed by an adult, in this case the teacher, who presents signs and meanings to the children in order to guarantee communication between them.

Vygotsky (2009) states that there are two types of concepts: spontaneous concepts and scientific concepts. The former are those that are formed as a result of the subject's interaction with the physical world in their daily lives. Scientific concepts are generally formed in school environments. The development of an everyday concept needs to reach a certain level for the scientific concept to be learnt by the child. Cavalcanti (2007) complements this reasoning by saying that children have knowledge when they arrive at school and it is on this prior knowledge that teachers need to plan their actions.

Vygotsky (2009) states that the difference between spontaneous concepts and scientific concepts represents differences in learning at school and outside of school. For this reason, Vygotsky developed two key concepts, the Zone of Actual Development (ZAD) and the Zone of Proximal Development (ZDP). The ideal denomination, according to Prestes (2010), would be the Potential Level of Development (NPD). This is because the activities carried out by children create developmental possibilities so that internal processes are awakened as well as their psychological functions.

The Potential Level of Development (PDD) encompasses the psychic functions where a human being's abilities develop. It is also a set of abilities that a human being can develop when assisted by someone more experienced, in the case of school, this assistance can be provided by the teacher or by classmates who have already managed to understand what has been taught. The skills that human beings can develop when assisted by someone can become clues as to where we can intervene so that the result is positive in the teaching and learning process (VIGOTSKI, 2009).

Development and learning are interlinked, even if a child's learning begins long before they reach school. For this learning, Vygotsky (2009) calls it real or effective development. When the child arrives at school, new elements are introduced that will form part of what he calls potential development, i.e. knowledge that is scientifically constructed.

It is through others, through the adult, that the child becomes involved in pedagogical actions. Everything in the child's behaviour is fused, rooted in the social. The child's relationships with reality are therefore social relationships from the outset. In this sense, the school is a universe that promotes artificial development, which restructures and influences the development of behavioural functions (VYGOTSKY, 1984).

The actions that children carry out independently, without anyone's help, are part of their development, and what they are able to do with someone's help is also part of their development. In order to carry out actions in the classroom, children need dialogue, collaboration, imitation and shared experimentation (REGO, 2011).

Representation, for Vygotsky (2000), is the emphasis placed on action that results in learning. This process occurs from the outside in and is full of historicity, as it was created by society (sociocultural genesis). In this way, we present the actions that guided the context of this research, which sought indications of how to teach Geography in the early years of Primary School I and understand the complexity in the formation of the basic concepts of School Geography, specifically referring to the topological notions of laterality and localisation. We chose to present the actions, through pedagogical activities and the way they took place, even if in a simple way, in the face of such a significant theory as Activity Theory.

1.2 Pedagogical actions in the teaching and learning relationship

This research is characterised by a case study, a way of researching teaching practice in the early years of basic education. According to Yin (2001), it is a strategy for answering "how" and "why" questions.

The subjects of this research were children enrolled in pre-school and 1st grade in the municipal public school system in the municipality of Itapejara **D'Oeste** (map 1) in 2016 and 2017. This case study made it possible to analyse children aged five (5) and six (6) from four (4) schools in the **municipality of Itapejara D'Oeste, totalling** 114 children, five (5) teachers and one (1) coordinator.

This dissertation presents pedagogical actions as a process to help teachers teach important geography concepts.

MAP 1: Location of the municipality of Itapejara D'Oeste.
Source: FRANCISCHETT, Mafalda Nesi, 2018.

The four (4) schools in the municipality of **Itapejara D'Oeste are:** 1) Escola Municipal Nereu Ramos, located at 601 Rui Barbosa Street, Centre; 2) Escola Municipal Irmão Josafat Kmita, located at 365 Canelinha Street, Bairro Industrial; 3) Escola Municipal Prof. Pedro Viriato Parigot Souza, located at 2555 Guarani Street, Bairro Guarani and; 4) Escola Municipal **do** Campo Valentim Biazussi, located at 2555 Francisco Salviato Parigot Souza Street, Bairro Guarani. Pedro Viriato Parigot de Souza Municipal School, located on the access road to Rua Guarani, 2555, Bairro Guarani, and; 4) Valentim Biazussi Municipal School, on Avenida Francisco Salvi, 903 (Barra Grande community). The latter has two multi-grade classes, one (1) with Pre-school/1st/2nd grade and the other with 3rd/4th/5th grade. The 3rd/4th/5th grade class is not included in this research.

One of the main characteristics of the case study is to analyse contemporary phenomena, inserted in real-life contexts, with the need to understand some complex social phenomena (YIN, 2001). Therefore, relating pedagogical actions that can help children form concepts that are important to Geography also makes a scientific contribution to educational processes.

> The case study relies on many of the techniques used in historical research, but adds two sources of evidence that are not usually included in a historian's repertoire: direct observation and a systematic series of interviews (YIN, 2001, p. 25).

A systematic series of interviews with teachers, students and coordinators was adopted, based on the pedagogical actions that contribute to the formation of Geography concepts (front/north, back/south, right/east, left/west).

The case study is intrinsic, but not singular, as it analyses actions that teachers carry out during the teaching process, with particular and in-depth attention to the pedagogical actions developed to improve learning in the classroom (YIN, 2001).

Four (4) pedagogical actions were selected for the research script. In between interviews, the teachers worked on new pedagogical actions. Whenever the children were able to relate to what was being asked of them, a new action was adopted as a methodology for analysing the children's understanding of the process of forming new concepts.

For Yin (2001), a case study takes into account the questions and theoretical propositions that form part of this context; what the units of analysis are; the logic that links the data to the propositions. Also the criteria for interpreting the findings, i.e. analysing the data.

In this text, the dissertation presents some actions that early years teachers have taken to promote children's knowledge. Analysing the actions of both the children and the teachers reveals steps towards clarifying situations where learning was more or less significant for the children.

Depending on the pedagogical action adopted by the teacher, the information may or may not have meaning for the child, and this can help them to transform their thinking (reconceptualisation), guiding teaching practice. Regarding the case study, Yin (2001) says that this methodology seeks to present actions that prioritise cooperation and promote experimentation in children.

This research looked at children between the ages of five (5) and six (6) because this is the age group in which they are leaving pre-school to start primary school. According to Duarte (2008), this is the phase that introduces the child to the world of education and pedagogical actions become methodological resources that help during this process.

It presents some of the actions that teachers in the early years worked on to promote the construction of knowledge. The research arose from the need to study the difficulties children have in forming important concepts for Geography such as: front/back/right/left and relating them to North/South/East/West.

The analysis of the teachers' actions reveals stages that favour changes in the way Geography is taught in the early years, in order to provide situations where learning is meaningful.

In a case study, the research goes through three stages. The first defines what is going to be studied, i.e. the unit of analysis and the social context. The second determines whether the study will be singular or multiple. In the third stage, theoretical development takes place, which helps to lend credibility to the data and organise the strategies for its analysis (DUARTE, 2008).

The research began in 2016 with 26 children aged five (5) with the aim of diagnosing whether they recognised the colours red/green/blue/yellow, as they were a reference for the sequence of future actions involving the formation of concepts such as: front/back/right and left.

At the beginning of 2017, the research involved 25 children aged six (6). The interviews were resumed, repeating the actions from the previous year's research script. By interviewing the children, it was possible to understand whether they related the pedagogical actions to the geographical concepts of North/South/East/West. This was a stage that required a longer sequence of actions.

The aim was to investigate teachers' actions to relate content to learning. The interviews with the children and their teachers were carried out individually and the work was structured around the colour variable, because it was believed to be a way for the children to better relate the information. At the school, the interview location was set up with a table and chair outside the classroom, near the window, so that the children could be close to their classmates and also to the teacher. This was done so that the children wouldn't feel embarrassed.

The teacher accompanied the child to the researcher. A different action was proposed at each moment of the interview. For example, when the child related the action to what was being proposed, they moved on.

In between interviews, the teachers worked on activities to help the children understand the proposed action. This intervention took place whenever there was any difficulty in understanding the concepts.

One of the reasons for defining the concepts of topological relationships was because the relationships that children develop in the family and outside it are part of a common interest. Children begin to relate objects, people and places to themselves. In this way, the teacher in the early years helps to articulate the actions worked on in the classroom and outside it, favouring the appropriation of the information necessary for the formation of concepts that are important for the development of abstract thought.

In primary school, geography teaching also faces limits if children are to be able to situate themselves in places beyond where they live (school/home/municipality/world). To this end, the teacher plays an important role in enabling children to develop their spatial knowledge. They need didactic opportunities, in a space for dialogue, interpretation and comparisons related to changes in the space they live in. This spatial knowledge is understood in topological relationships, through the school knowledge proposed in the Geography curriculum.

According to Brasil (2010/2017), the General National Curriculum Guidelines for Basic Education (DCNEB) and the National Education Guidelines and Bases Law (LDB) state that the knowledge produced in schools must, through public policies, produce culturally constructed knowledge and values. The challenge proposed in the Guidelines is to reduce or eliminate the gap between the different pedagogical proposals and the reality present in the classroom.

> [...] the common national base interacts with the diversified part, at the heart of the

> process of constituting the knowledge and values of children, young people and
> adults, highlighting the importance of the participation of all segments of the school
> in the process of drawing up the institution's proposal which must, under the terms
> of the law, use the diversified part to enrich and complement the common national
> base. (...) both the common national base and the diversified part are fundamental
> for the curriculum to make sense as a whole. (BRASIL, 2010/2017, p. 32).

Developing the ability to learn goes beyond mastering reading, writing and calculation. It is understood as the child's ability to learn the social and political meanings present in knowledge, which are transformed into new social codes. These social codes support the child's dialogue with the world. Geography content needs to help children understand the natural and social environment, as well as the political system, the economy and the use of technology, based on the rights that underpin society. Geography lessons need to develop children's ability to observe phenomena, problematise situations and analyse processes. They need to learn new attitudes and values through dialogue and the exchange of experiences, favouring bonds of human solidarity and mutual respect (BRASIL, 2010/2017).

Geographical knowledge produces basic guidelines for a curriculum, which selects and transforms content so that it can be taught and learnt. They also serve as elements responsible for the ethical and political formation of each child. The different areas of knowledge encourage communication between the different systematised knowledge (diversified disciplines) and between other knowledge (which the child brings from home). (BRASIL, 2010/2017).

In the actions worked on in this research, the colour variable was present because it is believed to be a way for children to relate the content worked on to the information present in the actions. This is because, according to Bertin (1986), the visual image helps the child to make relationships between the information in the activity and to relate to the differences in terms of order and colour. proportionality of the data, which follows a logical classification or sequencing, building a homogeneous and coherent system for analysis. The colours and their order present in the actions help to memorise the information, as the image is a variable that is perceived instantly. This helps to obtain the data and understand the information.

The information contained in the activities was placed in such a way as to show a matrix, which according to Bertin (1986), is responsible for ordering and providing the answers, because the property of the image is materialised by the child's gaze. The colour variable used gives meaning and promotes reflection, as it was drawn up based on information that follows the colour sequence: blue/green/yellow/red, facilitating graphic communication.

The simplicity of colour placement creates a repertoire of information **that "[...]** simplifies the discovery of the sought-after element". (BERTIN, 1986, p. 07). The visual separation of colours allows children to recognise the order of the colours, and then include new information relating the colours to a conceptual ordering a posteriori, which helps in the construction of characters and the overlapping of information.According to the action carried out below, the aim was to identify whether the children recognised the basic colours (Figure 1).

FIGURE 1: Colour identification.

Organisation: CHIAPETTI (2016).

The colours blue/green/yellow/red were distributed, as shown in figure 1, so that the child could relate the colour and order of each one in the position indicated. Colour and order are part of the sequence related to concepts. The purpose of differentiating the colours is to identify the positioning of the colours, preparing the children to identify and order them in preparation for the cardinal directions.

The use of colours in the proposed actions makes visual perception a pedagogical process for the development of perception, allowing children to concretise information and reflection, materialising notions that may be abstract. The use of the colour variable allows reflection and understanding of the issues that are relevant to this research (BERTIN, 1986).

> Graphic representation is one of the basic sign systems designed by the human mind to store, understand and communicate essential information. As a "language" for the eye, graphic representation benefits from its ubiquitous characteristics of visual perception (BERTIN, 1986, p.2).

The specificity of the actions proposed for this research is centred on graphic representation, as a main element of spatial perception through colours, which will be present from the beginning to the end. The proposed actions seek to associate the visual variable colour: green/yellow/blue/red with the spatial positions: front/back/right/left.

A timetable was drawn up (Table 1) with the actions carried out in 2016 and 2017. The first phase of the research was carried out in 2016 and 64% of the children recognised the colours; 36%

had some kind of difficulty differentiating between them, necessitating new pedagogical actions to help them recognise them.

In the second phase, also held in 2016, they worked on the topological relationships front, back, right, left, related to the colours worked on in the first phase green/yellow/blue/red. The same difficulties recurred. There was a need for new planned actions. The 2016 school year ended with two children having difficulties relating colours to topological relationships.

The third phase began in 2017 with the first graders and none of them had any difficulty identifying the colours. The activity used in the survey was the same as the previous year. As all the children were able to identify the colours, we moved on to the fourth phase, which sought to identify whether the children were able to relate green/yellow/blue/red to the concepts of front/back/right/left. Ten (10) children were unable to relate the colours to the concepts. However, the continuity of the work by the teacher and the researcher over eight (8) weeks meant that they were all able to make the connection.

In the fifth phase, 76% of the children identified the position of each colour. It took another three (3) weeks of work before they were able to relate all the colours to the position indicated.

The sixth phase of this research involved working with the geographical concepts of North/South/East/West. This phase required a different sequence of actions, as only 12% of the children related the concept worked on and 88% of the children had some kind of difficulty relating colour to the concept.

The table below shows the quantification of the actions, with the dates and the number of children who had some difficulty or not during the process.

TABLE 1: Schedule of actions carried out with the children

Phases	Day of action	Action	No difficulty	%	They had difficulty	%
First phase: a total of 26 children	04/07/2016	Colour identification	16	64%	10	36%
	12/07/2016		21	84%	5	16%
	04/08/2016		24	96%	2	4%
	12/08/2016		24	96%	2	4%
	19/08/2016		26	100%	0	0%
Second phase: a total of 24 children	03/10/2016	Identification of Right/Left/Front/Back	16	64%	10	36%
	10/10/2016		17	68%	9	32%
	17/10/2016		20	75%	6	25%
	24/10/2016		23	82%	3	18%
	31/10/2016		24	86%	2	14%
	25/11/2016		24	86%	2	14%
	02/12/2016		24	86%	2	14%
	05/12/2016		24	86%	2	14%
	12/12/2016		24	86%	2	14%
Third phase: a total of 25 children	17/03/2017	25	Colour identification	100%	0	0%
Fourth phase: a total of 25 students	27/03/2017	10	Identification of Right/Left/Front /Back	40%	15	60%
	10/04/2017	23		92%	2	8%
	24/04/2017	2		8%	0	0%
	03/05/2017		19	76%	6	24%

Fifth phase: a total of 25 children	08/05/2017	Identifying colours and their order.	25	100%	0	0%
Sixth phase: a total of 25 children	22/05/2017 29/05/2017 08/06/2017 12/06/2017	Identification of the concepts of North/South/East/West.	3 3 18 25	12% 12% 69% 100%	22 22 08 00	88% 88% 31% 0%

Organisation: CHIAPETTI, 2016/2017.

The actions in phase 1ª took place over a period of 45 days, following a weekly interview. During this period the actions were worked on by the teacher. It was noteworthy that two (2) children had to relate the colours to objects in their living space, as in the case of blue relating to the sky and yellow to the dried banana leaf on the side of the school. The other child who had difficulty differentiating the colours related yellow to the banners painted on the school. As for the colour green, both children related it to the trees in the schoolyard.

For the action **called** "Identifying the topological relationships of right/left and front/back" (Figure 9), nine (9) weeks were needed. The actions were also worked on by the teacher. Two (2) of the twenty-six (26) children had difficulty differentiating between right and left topological relationships. This information can be seen in Table 2.

The activity mediated by teachers, according to Moura (2010), favours its execution, since mediation consists of the activity of teaching through an action. In this sense, there is the thoughtful act of the child carrying out what they have been guided to do, through the mediating action of others (LEONTIEV, 1978).

Table 2 below shows that the greatest difficulty was in recognising the colour yellow - coincidence or not, the other colours are present in the school's paintwork. Perhaps this helped the children to identify the colours green/blue/red.

Table 2- Recognition of the colours blue/green/yellow/red (2016)

Colours	Total number of children who identified the colours	Hits (%)	No hits (%)
Blue	24	96%	4%
Green	24	96%	4%
Yellow	23	92%	8%
Red	24	96%	4%

Organisation: CHIAPETTI (2016).

The difficulty in differentiating the colours could be because, according to the teacher, no individual work had been done on colour recognition until then. When asked collectively, all the children were able to say the name of the colour they were being shown.

When the activity transforms the child's thinking, it contributes to the appropriation of knowledge. The teacher therefore has the task of organising in such a way that teaching becomes a pathway that favours learning and development. This happens when the child understands the reasons for their action, so interest and clarification of the object must coincide (MOURA, 2010).

Children learn when they understand the relationship between what they have produced and what has driven them to act (action). For learning to take place, it is necessary to provide the means for the child to be active in the production of knowledge (MOURA, 2010).

Therefore, pedagogical action needs to be aimed at a goal and driven by the will to do. In this way, the objectives will be achieved through the realisation of the children's actions in relation to the proposed activity. In this sense, the teacher has the main role of mediating the activities that are developed and transformed into knowledge for the child.

> The relationship between sign and meaning involves mental action that goes beyond the simple use of ideas and their associations, as the child begins to make use of the first concepts, identified as general (MOURA, 2010, p. 62).

It's worth remembering that interactive strategies are those that take into account the child's prior knowledge, their shared understanding in the classroom, the contextualisation of the content to be learned and the use of various examples (ALVES, 2007/2016).

Learning new content is the product of a constructivist mental activity carried out by the student, as well as a socio-interactive activity. The possibility of constructing a new meaning, of assimilating new content, necessarily involves the possibility of coming into contact with new knowledge. And the new knowledge is learnt by the subject from the moment they become aware of this reality (ALVES, 2007/2016, p. 42).

Learning is not the repetition of textbook discourse, but the result of the relationship of meaning and significance established between prior knowledge and the new content/concept (ALVES, 2007/2016). Thus, for Vygotsky (2001), in the relationship between thought and language, elements are presented that show meaning and sense, i.e. the sum of these elements. Therefore, meaning is a dynamic formation and significance is a part of this formation. Sense allows investigation, while meaning is specificity.

For children to appropriate knowledge, the actions they work on need to be appropriate to the objective they want to achieve. The child's experiences play an important role in the process of building knowledge and occur through the practical and verbal relationships between the child, the teacher and the other people who are part of the school environment (MONTEIRO; GHEDIN; KRUGER, 2017).

Children don't all learn in the same way and the environment interferes with learning. In order to learn **concepts, "[...] the** child **needs to** develop appropriate actions, and these actions are first external because they take place with the guidance of the adult and then with internal actions." (MONTEIRO; GHEDIN; KRUGER, 2017, p. 04).

Therefore, if the child receives social help and educational assistance that stimulates them, the key point will be a shift from the main activity to meaningful learning (MONTEIRO; GHEDIN; KRUGER, 2017). This will bring about a cognitive transformation that will lead to the appropriation of new knowledge.

Even if the child is interacting with others their own age, the teacher needs to be present, playing their role as mediator of this relationship between classmates. They need to be made aware of their responsibility when carrying out the requested activities (MONTEIRO; GHEDIN; KRUGER, 2017).

At pre-school age, which is between five (5) and six (6) years old, children learn through play and the teacher mediates the games as proposed. The teacher needs to be clear that "[...] **what arouses their motivation is** conflict. [...]" **(MONTEIRO; GHEDIN; KROGER, 2017, p. 06).**

The teacher, as a mediator, develops pedagogical means that enable children's reasoning and way of thinking, respecting their differences and their own development (DUARTE, 2004).

> [...] the teacher is responsible for mediating pedagogical activity, since its purpose is to guarantee the appropriation of the material and cultural goods of humanity that have been systematically elaborated. For this reason, he emphasises that understanding the social meaning of pedagogical activity will allow us to know what motivates the practice of teaching. This concern is due to the meaning and sense that the teacher gives to a given piece of content, as well as the influences he or she suffers and which can interfere with the quality of the teaching activity. Therefore, systematised teaching introduces study as an activity for the student so that they can appropriate the knowledge produced by humanity, where internal and external needs motivate the child's interests, but it is up to the teacher to use the main activity, which in the pre-school phase is play or games. (MONTEIRO; GHEDIN; KRUGER, 2017, p. 10).

The development of the child's intellectual activity requires the teacher to be clear about the processes that lead to the formation of mental operations. **As such, "[...] the teacher needs to make their lessons viable** through concrete materials and [...] the actions that the students must carry out must be guided [...]" (MONTEIRO; GHEDIN; KRUGER, 2017, p. 11).

> Another way of working with activity theory is through the problem-solving technique, which is a pedagogical resource for problematising situations that are part of the student's daily life on social issues that have a cultural, political and psychological basis that directly or indirectly affect their life. With this, the teacher will be stimulating reflective thinking and arousing interest in the search for a solution (HENNIG, 1998), in this way, the student will appropriate the processes of objectification from intellectual activities. (MONTEIRO; GHEDIN; KRUGER, 2017, p. 11).

Based on the report, **"[...] the pedagogical activity should converge towards the** objective of the main activity [...] exploiting the potential of both activities and giving them a meaningful character". (MONTEIRO; GHEDIN;

KRUGER, 2017, p. 11).

Therefore, pedagogical action, as mediation, acts on the content and needs to help the child overcome the difficulties presented in previous stages of the teaching/learning relationship, through new methodologies (MONTEIRO; GHEDIN; KRUGER, 2017).

For knowledge to occur, children need to interact with information. The construction of knowledge takes place gradually and unevenly. That's why it's necessary to use individual work to diagnose difficulties.

It was in this sense that the teacher resumed her investigative work with the action, using geometric figures and colours, using new strategies to enable the children to identify the four colours (green/yellow/blue/red). According to the teacher, the action involved different subjects and this was a way of relating mathematical content (geometric shapes) to colours.

The actions worked on need to be thought of as a body of knowledge full of values, knowledge and procedures. The teacher needs to bridge the gap between the different cultures presented by the children and scientific knowledge so that learning is meaningful (CAVALCANTTI, 2015).

She also continued with the content proposed in the school's Curriculum Guidelines and deepened the explanations related to colours and shapes, as shown in Figure 2:

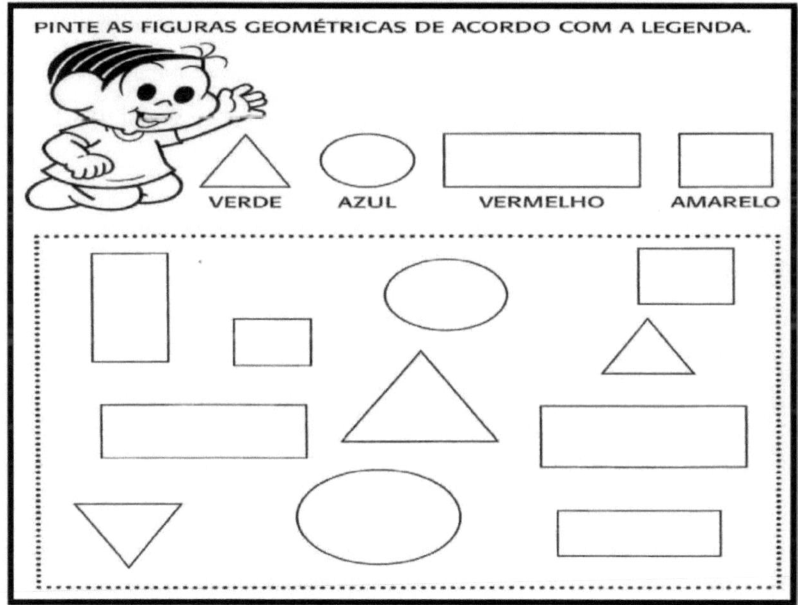

FIGURE 2: Geometric figures and colours

Source: ESMERALDO, 2010, p. 67.

The teacher's work with the children enabled them to see different geometric figures and their relationship with colours, which is proposed in the subject of Geography at this stage of education. This universe of colours and shapes enables learning through exploration and contact with visual differences. The action required attention to the legend and the specific colours for each geometric shape.

> To mediate this, he relies on school culture, the body of knowledge systematised in science, in this case geography, and structured pedagogically to make up the knowledge necessary for the general education of citizens (CAVALCANTI, 2015, p. 72).

Geography teaching contributes to the development of children's thinking about the world and reality. This depends on the didactic approach to the content and the actions that become didactic resources for thinking about geographical space. For this reason, the geography taught needs to be integrated with the child's culture, in their daily lives. This encounter results in meaningful learning, which broadens culture and promotes a new view of facts (CAVALCANTI, 2015).

Pedagogical action needs to become a symbolic mediation of what is real; action therefore needs to give meaning to the content to be taught to children (CAVALCANTI, 2015).

> The teacher's task of deciding how to structure and select school content in order to carry out constructivist-based teaching presupposes a certain understanding of some questions such as: what does it mean to give meaning to the content presented to children? What needs to be done for this process of assigning meaning to take place? What content favours this activity of assigning meaning? In short, what does it mean to work with meaningful content in Geography teaching? (CAVALCANTI, 2015, p. 73).

Understanding reality is not a simple matter, and we can't think that contact with reality alone can provoke awareness in children. What is important is the way in which the teacher conducts his or her practice, which provokes a break with the old way of seeing things (ALDEROQUI, 2002).

To make the action more engaging and participatory, the teacher combined the painting activity with the use of gouache paint. This time she added new colours, orange and purple, as three of the children had difficulty differentiating between the colours they had been taught. This involved all the children. Those who differentiated helped those who still had difficulties.

For this action, the teacher handed out the printed action and the children coloured the first three circles with the indicated colours (blue/yellow/red) using gouache paint. Then the children mixed colours, yellow and red resulting in orange, yellow with blue resulting in green and red with blue resulting in purple. According to the teacher, this action provided moments of interaction and joy among the children who showed enthusiasm every time a new colour appeared (mixing colours).

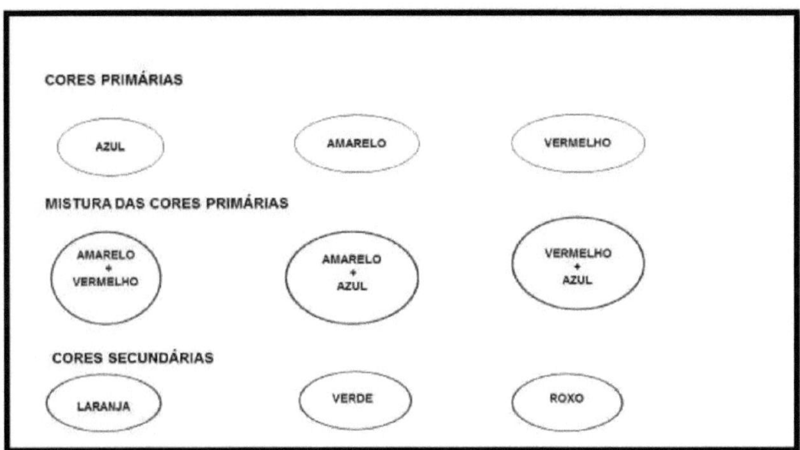

FIGURE 3: Adding colours

Source: Teacher's personal archive

The mixture that will result in new colours really caught the children's attention. They discovered that they can create other colours to colour in. Mixing primary colours to create secondary colours aroused the children's curiosity.

Learning is directly linked to cognitive action that requires a series of transformations, allowing the child to reconstruct the object studied. For this reason, there is a need for co-operation between the children and the mediating role of the teacher to trigger action in the child (GUERRERO, 2007).

> The main elements of learning activities consist of: Apportionment of initial actions and operations; Exchange of modes of action based on the introduction of different models of action as a means of common transformation of the model; Mutual understanding, favouring the achievement of relationships between the action itself, its result and the actions of one of the participants in relation to another; Communication, ensuring exchange and mutual understanding; Planning of individual actions; Reflection to go beyond the limits of actions.
>
> (GUERRERO, 2007, p. 122).

According to Guerrero (2007), starting with sensory and material actions with a symbolic, graphic basis allows children to mentally construct the representation of colours. In this way, the child is taken from the social to the individual to reach interpsychic action and then intrapsychic action. That's why collective as well as individual action is so important.

The action that is structured in teaching allows it to be mediated by the content and its purpose is the collective solution of problem situations. It becomes guiding when it presents essential elements for educational action, respecting the dynamics of interactions. However, they don't always achieve the result expected by the teacher who sets objectives, defines actions and chooses instruments that can help in this process, which is why interaction allows for meanings shared by the children that change during discussions.

However, all actions that contribute to children's learning are important, because solving problem situations triggers the teaching and learning process, which allows for the reconstruction of concepts and the reformulation of representations capable of teaching them how to learn.

The result of the colour identification action was discussed with the teacher. It was concluded that new actions should be introduced in the classroom. For this reason, she worked for one (1) week with objects in the classroom that involved the relationships of front and back. The children coloured using the blue/green/yellow/red colour sequence. However, in this action, the teacher added other new colours (pink and orange) so that the other children would also have the opportunity to continue the knowledge they had acquired.

> The organisation of learning activities capable of promoting not only the acquisition of specific knowledge but also the cognitive, motor, social, cultural and affective development of children is one of the duties of teaching know-how. These activities are related to all the content and concepts studied in the classroom and aim to respond to the act of teaching, which, in turn, responds to the needs of the student seeking to learn (GUERRERO, 2007, p. 124).

Pedagogical action needs to be meaningful for the child. If it is meaningful, it promotes learning. For Vasconcellos (2005), when working with knowledge there is a process of appropriation and construction of the knowledge itself, which he defines as the content/methodology relationship.

> We are experiencing a temporality in which the centrality of the student in the learning process places us, [...] as researchers and investigators of who they are, the ways of being, living and learning of the bodies we seek to teach. (GIORDANI, 2015, p. 152).

When planning this action, the teacher tried to keep her objectives clear, i.e. to get the children to differentiate between the colours green/yellow/blue/red, as shown below:

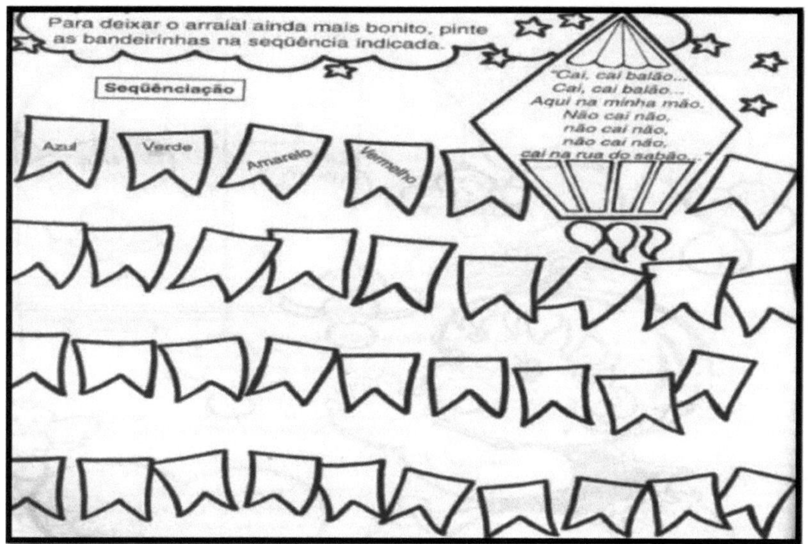

FIGURE 4: Colour sequence

Fonte:http://tersolbsm65.pbworks.com/w/page/20593398/Espa%C3%A7o%20e%20Forma%20a tiv%201-2-3-4-5-6-7. Accessed on 29 January 2017.

This action aimed to get the children to make the connection between the colours and the sequence they had to follow. The children were shown a box with different colours of crayons, and this novelty caught their attention. The aim was to help the children differentiate between the colours. This also happened with the help of their classmates, as the teacher asked the children to sit in a group, although some preferred to do it individually, which was respected. Collectively, all the children managed to carry out the action successfully. The sequencing of colours in the action develops logical reasoning, as well as promoting colour identification. The action helped the children in the process of naming colours and helped develop their observation skills.

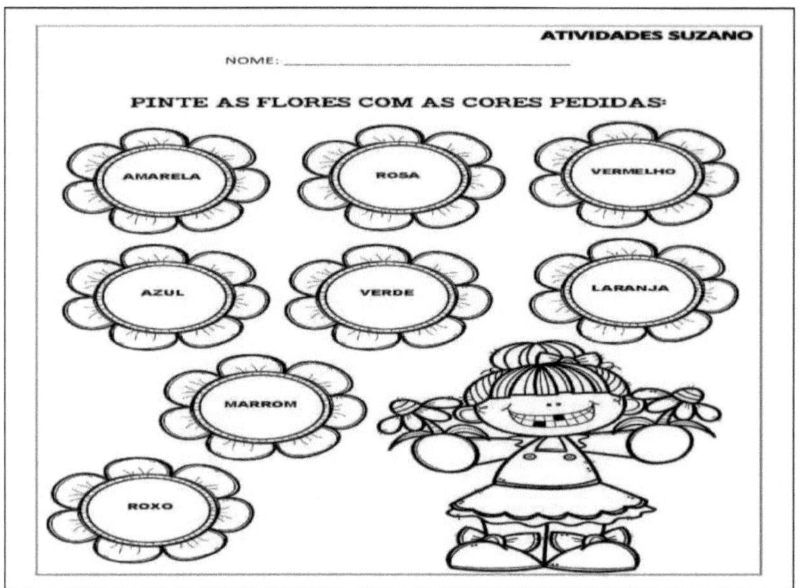

FIGURE 5: Adding new colours

Source: http://www.atividadespedagogicasuzano.com.br/2016/09/letras-e-cores.html. Accessed on 27 May 2017.

The teacher realised that some children were still having difficulties. That's why she enlisted the help of the coordinator, who helped her select new actions.

After one (1) week, the investigation was repeated, this time only with the three children who hadn't managed to identify the colours. The result was that one (1) of the three (3) children only recognised the colour yellow, and did not recognise green, blue and red; while another recognised the colours yellow and red, but had difficulty recognising green and blue, and the third child continued to have difficulty identifying the four (4) colours.

When asked, "Why did you think the children couldn't identify the colours?", the teacher said she thought **it** was because "They forgot and didn't understand the difference between them". The teacher said she thought it was because *"they forget and don't realise the difference between them",* **while the** pedagogical coordinator replied that: Perhaps *because of a lack of stimulation in the environment where they live, there is probably some **learning** difficulty **with these children"**.*

They were also asked if this was a common occurrence, and both said no: *"Most children at this age already recognise colours"* **(teacher),** *"**Because** most five (5) year olds who attend pre-school already recognise colours. Not identifying colours can happen to children who already have some difficulties"* **(coordinator).** We were also asked: "What was/is the main problem why the children didn't identify the colours?". **The teacher replied:** *"Each child has time **to develop** their* **ability".** While the coordinator **said:** "The *children learn the colours around the age of two (2), when* *the child has learning problems they learn the colours easily, only when there is a problem the children often take a long time to identify **the colours.** "*

The action worked on by the teacher made a positive contribution to the process of recognising colours. However, it was still necessary to come up with new actions, as some children still didn't recognise certain colours. Based on what the teacher had said, other learning situations had to be organised. According to Guerrero (2007), promoting integration activities among the other members of the school community allows them to share and carry out tasks that trigger collective knowledge.

When she presented the diagnosis to the teacher, she understood the need to work specifically with the colours blue and green, because it was with these two colours that the children still had difficulties.

> The organisation of teaching to promote learning is a premise of Didactics. This means that one of the duties of teachers is to organise teaching based on the constituent elements of the activity of teaching. These elements include social relations and the historical context in which they occur.
> Being a teacher implies mastering specific knowledge of certain types of knowledge produced and considered socially and culturally relevant. This knowledge constitutes true cultural heritage, because through it we can understand the origin of social or physical phenomena and build objects that are useful to society (GUERRERO, 2007, p. 124).

The teacher had to choose and plan actions to help children with difficulties in the process of constructing knowledge in relation to the colours blue and green.

Guerrero (2007) says that the way teachers transmit knowledge is a way of relating to and organising children's learning and also a means of devising teaching strategies.

Geography develops its own language, the geographical language, which triggers in children a process responsible for reading the world, a spatial view of things and elements. This spatial view is permeated with concepts, which are important requirements for children to analyse phenomena from a geographical point of view. When children are able to incorporate this language, they are able to operate rationally with some of Geography's own concepts (CAVALCANTI, 2015).

In each piece of content presented to the children, there was an introduction to geographical concepts and the action proposed by the teacher potentiates and enables the understanding of these concepts.

The child is a being who needs to be considered in their entirety and not as someone who needs to be at the disposal of the teacher or the school. They need to be seen as individuals who live in society and occupy their place, and therefore have their own space. Children need to be given the opportunity to construct their own knowledge, and for this the teacher needs to develop an attitude of questioning, provocation and openness to enquiry. Provoking children's wonder (CALLAI; CALLAI, 1996/2017).

In view of the children's difficulty in differentiating between the colours blue and green, the teacher made new relationships between objects and colours by taking them for a walk around the school grounds, with the aim of identifying objects with the colours blue and green respectively. To reinforce the learning, the teacher selected an action (figure 6) with the colours blue and green, as can be seen below:

FIGURE 6: Topological relationships and colours

Source: http://www.smartkids.com.br/colorir/desenho-nocoes-espaciais-em-cima-embaixo. Accessed on 27 May 2016.

In addition to the colours, the teacher worked on the positions "above and below" (the children went on holiday for fifteen (15) days). When they returned to class, she worked with the children who hadn't managed to identify all the colours proposed. The result was that two (2) of them managed to differentiate between the four colours, while one (1) child continued to have difficulty with the colours blue and green.

Next, the teacher worked with actions involving the colours blue, red and yellow. With the representation of some vowels and numbers up to three (3). The action was with the sequencing of the colours blue, yellow and red, as shown in the figure below:

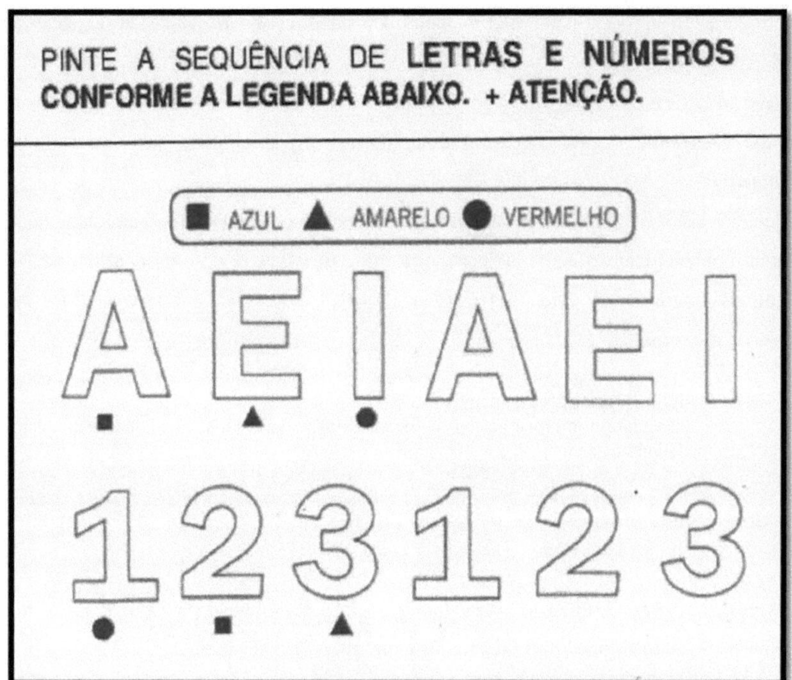

FIGURE 7: Sequence of letters and numbers

Source: http://pt.slideshare.net/blogdatiaiolanda/apostilha-da-educao-infantil-nvel-i-e-ii. Accessed on 29 January 2017.
The aim of this activity was to enable children to relate numbers to colours by using the legend, as well as to differentiate between colours. It's an initiation into map reading. The aim was also to awaken interest in **images, exploring visual language in order to "[...] develop the** ability to read and communicate orally and in writing [...] and thus enable students to perceive and master space". (SIMIELLI, 2007, p. 98).

Children's participation in the activities carried out at school promotes reflection on how space is formed and organised, enabling them to understand the relationship between the elements that make it up (SIMIELLI, 2007).

Proportion, scale, laterality, reference and orientation are geographical concepts that contribute to demystifying cartography because the information is understood and not simply reproduced (SIMIELLI, 2007).

According to Simielli (2007), the complexity with which we present drawings to children increases according to the level of content of the school year attended. The activities are references for understanding cartographic representations and the legend facilitates understanding from the concrete (concept) to what is abstract (represented).

The structuring of a legend can be initiated by elements present in the activities. In this way, children are introduced to a process that broadens their understanding of cartographic representation and the legend (SIMIELLI, 2007).

Forty-five (45) days after the first and last surveys, the colour identification action was repeated for the fifth time (figure 1). All the children were able to identify the colours yellow/green/blue/red.

Knowledge is not something that occurs in isolation, as it is constructed through interactions between different contexts and different subjects. According to Alves (2016), by considering the children's intentions, interests and curiosity, the teacher enabled mechanisms to be created for the process of constructing meaning.

> [...] knowledge does not consist of a faithful representation of reality, independent of life experiences and joint actions, nor does it consist of a mere process of transmitting information, but is the result of negotiated, situated and contextually delimited action.
> In fact, there is a marked difference between a lesson that is able to mobilise students, that arouses in them the desire and curiosity to learn, and a lesson whose strategies disregard students' knowledge and even reject it. (ALVES, 2007/2016, p. 56).

For Vygotsky (1987), children are born into a social environment (family). It is in this environment that they establish their first interaction with others, in this case, the people they live with. During these interactions, there is mediation with adults who make them project themselves as subjects, and this is reflected in the school environment through the role of the teacher.

Vygotskyi (1989) sees the child as an interactive subject capable of developing knowledge when mediated by the other, in this case the teacher and the researcher. When children interact with other children or with the teacher, they create new ways of thinking. During the mediation process, the teacher can use different tools in the actions, so that the children can make increasingly complex interventions. The exchange that takes place when actions are carried out in the classroom results in a teaching and learning process, because it involves those who learn and those who teach, as well as the relationships that are built up between the two. The author encourages actions to be carried out collectively because of the co-operation and collaboration that children adopt when carrying out an action in partnership, as individually the result may not be as satisfactory.

In order for the children to appropriate knowledge, the teacher had to select activities that were appropriate to the content. In this way, the child learns what is being presented. For this reason, prior planning is necessary to define the activity needed to stimulate the teaching and learning process (MONTEIRO; GHEDIN; KRUGER, 2017).

In order to learn, children need to carry out appropriate, guided actions, **so we can consider "[...] changing the main activity as a** condition for meaningful learning". (MONTEIRO; GHEDIN; KRUGER, 2017, p. 04).

Children only learn when they actively participate in the educational process (MORETTI; ASBAHR; RIGON, 2011). In this way, they make sense of what they are doing, which motivates them to carry out the requested activities (DUARTE, 2002).

The repetition of certain pedagogical actions provides the support that guarantees stability and security for learning to take place. Children are subjects in the construction of their own knowledge. Therefore, as the mediator of this process, the teacher adapts his or her actions according to the development of the children's autonomy, providing new opportunities for repetition (DUARTE, 2002).

For Vygotsky (2004), the process of transmitting and appropriating knowledge forms the basis of the educational process. Therefore, for teaching and learning to be successful, certain elements must be present, including observation, imitation, execution and repetition. These are fundamental attitudes for knowledge to take place.

Observing, imitating, executing and repeating are not harmful activities when they involve active behaviour on the part of the child. However, when these elements lead to robotic attitudes, they can be harmful to the teaching and learning process. When imitation and repetition are carried out without conscious action, we run the risk of turning our educational practices into training (VYGOTSKY, 2004).

The help of the teacher or peers gives the child the opportunity to understand what is being proposed. The path that the child's thinking produces to relate old and new knowledge helps to form the concept, which is when the child appropriates and understands the new words. However, this process takes place gradually, as it involves the integration of new ideas in order to arrive at a conclusive fact, i.e. the formation of the concept (VYGOTSKY, 1984).

When a child imitates, they are already capable of understanding. They can intuit its meaning or sense. When a child imitates, it is because the imitated content is already accessible to their level of development, even if they cannot yet realise it independently and need someone's help or mediation. This is why imitation or repetition must be seen as creative activities when associated with conscious attention (VYGOTSKY, 2004).

1.3 Notions   of topological directions

The aim of this sequence is to relate colours to the concepts of right/left/front/back. When carrying out the diagnosis with the children to identify their prior knowledge, the reference was the body of the **"boy"** in the drawing. This is the beginning of the subject's inclusion in space through representation.

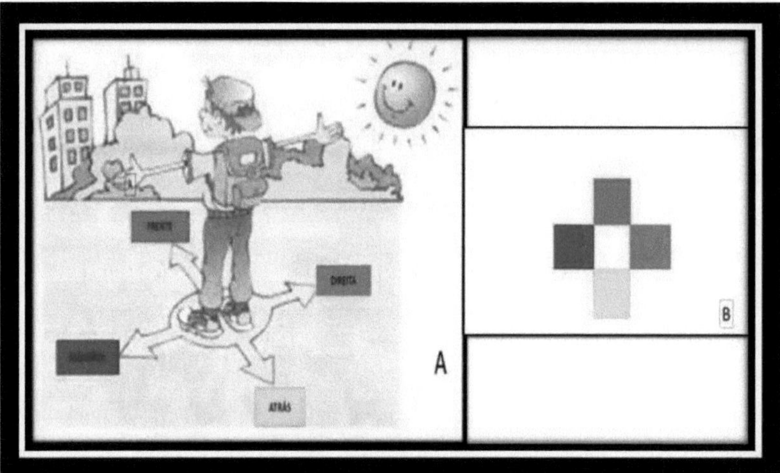

FIGURE 8: Identification of colours/topological relationships

Organisation: CHIAPETTI (2016).

The positioning of the colours has been maintained: the colour green, as can be seen in the images, is related to the front position (it is in front of **the "boy"),** the colour yellow (seen **behind** the **"boy"),** the colour red to the left **(left of the "boy")** and successively the colour blue to the right.

> The information contained in an image or cartographic representation depends on the variety or number of messages that can be understood, covered by the code that represents them, and depends on understanding the meaning of the messages (FRANCISCHETT, 2012, p. 93).

The "boy", represented in the figure below, is the reference for the children to position themselves and read the representation. Based on the representation, the children were introduced to the world of geographical representations in order to prepare the cardinal positions.

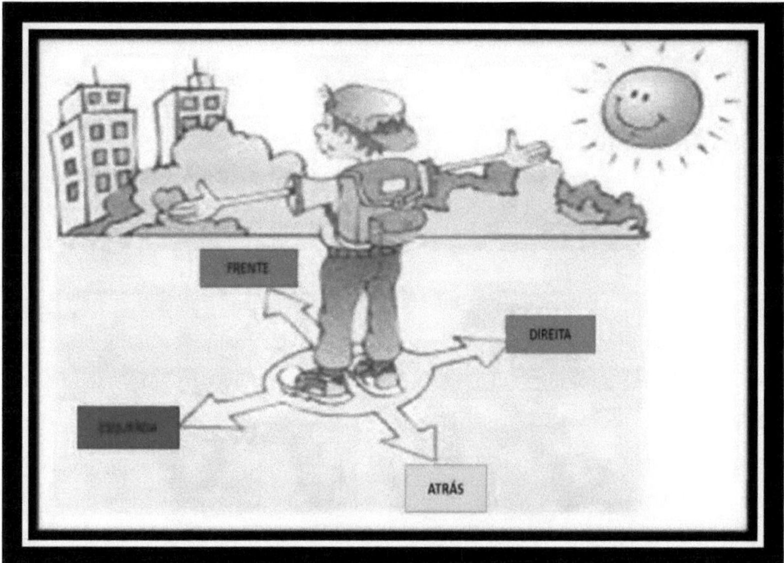

FIGURE 9: Identifying the topological relationships of Right/Left and Front/Back

Source: Available at <http://www.todoestudo.com.br/geografia/pontos-cardeais. Accessed on 20 October 2016. Adapted by CHIAPETTI (2016).

In the image, the topological relationships of right/left/front/back have the **"boy"** as their reference. The right and left directions are identified by the outstretched arms, while the front and back directions are related to the position of the **"boy's"** body in relation to the objects in the image, as well as the positions of the arrows and colours on the ground (FRANCISCHETT, 2009).

In order for pre-school children to understand the concept through representations, according to Jodelet (1989), there must be the activity

cognitive. For this to happen, the child, who is a subject in the process of constructing knowledge, needs to understand the representation in two dimensions. The first dimension is contextual: the child is in a situation of social interaction by means of stimuli (actions presented by the teacher and the researcher); the second is belonging: the child is able to perceive themselves as a social being and thus intervene in the development of the representation, promoting the internalisation of ideas, values and models that are common in the group to which they belong, in this case the classroom.The diagnosis showed that out of a total of 26 children interviewed, some were unable to identify the topological relationships researched. The data is shown in the table below:

Table 3: Topological relationships of right/left and front/back

Topological relationships	They succeeded	Percentage	They had difficulties	Percentage

Behind	24	92,30%	2	7,70%
Front	23	88,46%	3	11,54%
Right	23	88,46%	3	11,54%
Left	23	88,46%	3	11,54%

Source: CHIAPETTI (2016).

The five (5) year olds' greatest difficulty centred on understanding right, left and forward directions: three (3) children had difficulties. Two (2) children had difficulties with the back direction.

The appropriation of knowledge by the child involves the internalisation of psychic functions that are built up over the course of the school days: "Internalisation takes place in a complex network of interrelationships that **articulate the social activity of individuals".** (SIRGADO, 2000, p. 38).

On the importance of mediation, Oliveira (1988) emphasises that it is the process of an intermediary element intervening in a relationship; the relationship then ceases to be direct and becomes mediated by this element.

> The process of mediation, through instruments and signs, is fundamental to the development of higher psychological functions, distinguishing man from other animals. Mediation is an essential process for making voluntary, intentional psychological activities possible, controlled by the individual himself (OLIVEIRA, 1988, p. 33).

In a "[...] **broader sense, mediation is any** intervention by a third party that enables interaction between the terms of **a relationship".** (SIRGADO, 2000, p. 38). So, if human action takes place through mediation, learning occurs when there is interaction with the other, this social interaction between child/teacher, child/child and teacher/child, causes words or actions to be used as a means of communication and interaction. (MARTINS; MOSER, 2012).The role of mediation is that of an appropriate instrument that makes progress in thought, in which the psychic functions are, through the processes that, in the case of this research, take place within the school environment. In this way, the teacher's action implies a production of the content proposed to the children and a production by the child themselves, who internalises what they have understood (SIRGADO, 2000). Once these difficulties have been identified, the teacher continues to work:

FIGURE 10: Topological relationship of right and left

Source: http://artedeensinar10.blogspot.com.br/. Accessed on 29 January 2017.
The aim of this activity was to encourage children to differentiate between right and left, because these were the directions in which they had the most difficulty. This can be seen in table 4.

The teacher worked collectively in the classroom, indicating the directions with her hands. She explained the relationship with the sides, as she was facing the children and explained that her position would be reversed, that the teacher's right side would be opposite the child's side. The same goes for the left side. The teacher listed objects in the classroom belonging to both sides, such as the windows and the door, which are on the left side of the room, and the cupboards, which are on the right side of the desks.

When children can't differentiate between left and right, it's also difficult for them to orientate themselves in space and it's also impossible to understand the content of an image or representation (FRANCISCHETT, 2009).

In the thinking of pre-school children, pseudo-concepts are widespread because the meaning of words does not occur freely or spontaneously, the meanings of words are previously established by an adult and the child does not immediately assimilate an adult's way of thinking (VIGOTSKI, 2009).

> Man is the only cultural animal; because of his own characteristics, he develops ways of solving problems, conceptions of the world, arts, which are assimilated by new generations, whether to facilitate survival, to find the meaning of things or even because of a less immediate need [...] (VASCONCELLOS, 2005, p. 13).

The difficulty in differentiating right and left is greater than in differentiating front and back, because they have more plausible references for the children, such as the face and back, while right and left are limited to the sides of the body, with no other relationship.As the children continued to have difficulty differentiating right from left, the teacher worked on the following activity:

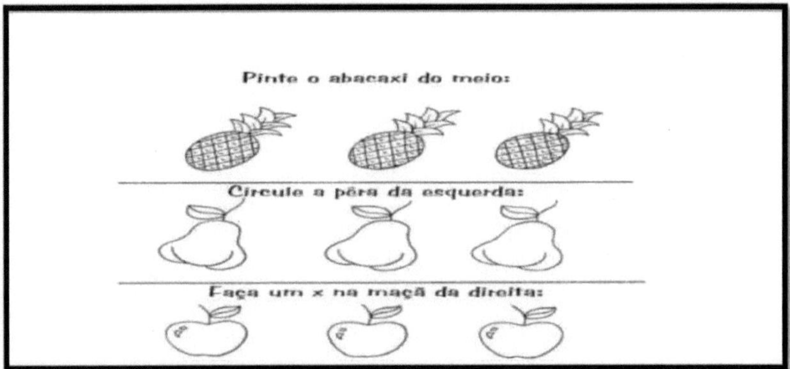

FIGURE 11: Right/left topological relationship
Source: http://pt.slideshare.net/LuceliaFialho/atividades-52183213. Accessed on 29 January 2017.

The work period, which lasted one (1) week, was one hour/class per week. The activity shown in figure 12 was carried out by the researcher during Geography lessons and involved the relations of right and left, with the **game of "Dead/Live".** A line was improvised with a rope. The letter D represented the right side and the letter E the left side. Under the mediation of the researcher, the children jumped in the direction requested. With this action it was possible to see that the children were able to identify the directions.

FIGURE 12: Right/left

Source: CHIAPETTI (2016).

During this action, the children managed to successfully carry out all the steps requested. It's worth remembering that the results are different from those achieved individually. This is because, as Guerrero (2007) points out, collective activities put children in the position of learners, but develop teaching attitudes in them, and this attitude organises teaching and favours learning.

> School content is characterised as a synthesis produced by social groups when dealing with problems arising from physical or psychological needs, the solutions to which improve the quality of life. Once chosen at a given historical moment by social groups that consider them relevant, they are conveyed at school in such a way as to allow new subjects to integrate into the dynamics of the society of which they are a part. Because they are cultural heritage, these contents also have a dimension related to the history of humanity itself, of an interdisciplinary nature (GUERRERO, 2007, p. 124).

Of the children who had difficulty relating right/left/front/back, two (2) were still unable to differentiate between the right and left sides. Recognising one's own body and relating it to the place it occupies is very important for children's learning. These were the same children who had difficulty performing the first action proposed at the start of the research. The data is shown in the table below:

Table 4: Demonstration of data on topological relationships in 2016

Topological relationships	Hits	Percentage of hits	Errors	Percentage of errors
Behind	26	100%	0	0%
Front	26	100%	0	0%
Right	24	92,30%	2	7,70%
Left	24	92,30%	2	7,70%

Source: CHIAPETTI (2016).

Identifying right and left, as already mentioned, is more difficult for the children because there is no reference they can make. The teacher worked on other actions with the class in an attempt to help the children who still had difficulties, such as the following activity:

FIGURE 13: Expanding the context of left/right topological relationships

Source:http;//espacoaprendente.blogspot.com.br/2014/04/atividades-para-trabalhar- lateralidade.html. Accessed on 29 January 2017.

The teacher worked on this collectively. She showed with her arm where the left was and then asked the children to raise their left arm and say what they saw on the left side of the room. She referred to the left side of the room with the presence of the wall and the objects. She then asked them to look at what was on the left. He then raised his right arm and asked the children to raise their right arm as well, identifying what was on the right. This time the reference was the door to the room, which was on the right. They then identified by colouring the boy who was indicated by an arrow.

The teacher worked on the directions of right and left orally, using classmates as a reference. Example: Who is to the right of this child? Which objects are on the right side of the room? What's the difference between the child's right and the teacher's right? In this sense, Francischett (2009) points out that the dialogue between the teacher and the children must permeate the communication between the reader and the signs presented in the activity, as well as their meaning in relation to the reality represented in it. **However,** "[...] the image, although compared to spoken language, is fundamentally different from it in that it can neither affirm nor deny anything [...]" (FRANCISCHETT, 2009, p. 07).

This gave the children moments of interaction, because when one child couldn't answer, the others helped. However, it's important to remember that collectively, the results may not be indicative of specificities, as children who have difficulties wait for the others to answer before responding.

Next, the teacher worked on right and left with gouache paint. The children divided the sulphite sheet into two parts, each labelled right/left. Each child depicted objects in the classroom that were on the same side of the sheet, for example: the right side of the sheet corresponded to the right side of the room and the children had to illustrate what was on this side.

The teacher also worked with the song **"Job's Slaves", in which** the rhythm and direction had to be followed in order to harmonise gestures and directions. At this point, according to the teacher, it was possible to identify which children had difficulty differentiating left from right, because from time to time the sequence of the game was broken by some children going in the opposite direction.

In this action, in order to differentiate right and left, at the signal everyone had to move to the right or left as requested. The two children continued to have difficulties performing the movements.

The environment provided by the teacher during the process of mediation between the child and knowledge has repercussions on the form of communication and interaction between teaching and learning (MARTINS; MOSER, 2012). Leontiev (1978) emphasises that in order for knowledge to be appropriated, the activity must correspond to the desired phenomenon, i.e. learning.

> [...] The use of artificial means - the transition to mediated activity - fundamentally changes all psychological operations, just as the use of instruments expands in an unlimited way the range of activities within which new psychological functions can operate. In this context, we can use higher logic, or higher behaviour with reference to the combination of instrument and sign in psychological activity (VIGOTSKII, 1998, p. 73).

The use of pedagogical actions reinforces children's learning and helps them interact with information, promoting the appropriation of knowledge.

> The fact is that no conscious act can exist without inner speech, without words, intonations and evaluations, every conscious act is already a social act, an act of communication. Speaker, listener and subject are constant participants in the communication of the creative event. Seeking to understand the dialectical relationship of material, form and content in artistic expression, Bakhtin concluded that, through the mediation of artistic form, the creator assumes an active position with regard to the content in conjunction with the material. (FRANCISCHETT, 2009, p. 07- 08).

The teacher continued with the actions, interspersed with the regular content.
We re-interviewed the two children who had difficulty differentiating right from left, and after 62 days the difficulty remained. The two (2) children ended the 2016 school year unable to differentiate right from left.

> It is not enough to be certain of the elementary perception of each sign in isolation, but one must prioritise reading at the level of the whole [...], the purpose of which is the reader's cognitive understanding (FRANCISCHETT, 2009, p. 08).

Vygotsky (2000) points out that there needs to be interaction between the child's language and their social world, because during the learning process (internalisation/expansion of cognitive competence), the other person's speech is very important, because the mediator's speech is a fundamental activity

for the construction of knowledge. It is through dialogue that children are able to build their world and their knowledge.

The knowledge produced by the child allows changes to occur in their cognitive capacities, because through experiences, which arise from socialising with their peers and teacher, the child develops skills and forms concepts. The repetition of actions awakens the child's ability to predict and deduce, which helps them reach their conclusions (VYGOTSKY, 2000).

With the repetition of actions that involve dialogue and the interaction of information, Vygotsky (2000) points out that the child may be able to do what he or she would previously have been unable to do without the help of science or even school interactions.

> [...] a concept is more than the sum of certain associations formed by memory; it is more than a simple mental habit, it is a real and complex act of thought that cannot be taught through training (VYGOTSKY, 2000, p. 104).

For Vygotsky (2000), knowledge is not something that happens in isolation, as it is a construction based on interaction and context. The interactive strategies carried out through different actions need to take into account the children's prior knowledge, objectives and expectations. However, we must remember that the collective construction of knowledge (dialogue and interaction) contributes significantly to learning.

In order for interaction to take place, Vygotsky (2000) states that there must be intentions on the part of the speakers, that they must be recognised by the listener. This process needs to be contextualised so that the child can build schemas that help them construct their knowledge, and this interaction can take place verbally or non-verbally. However, the interaction between the child and the information needs to be contextualised. Teaching and learning are actions that contribute to cognitive development when both (teacher and child) participate through experiences that promote the construction of new knowledge.

The knowledge constructed by the child in the classroom is the result of interactions in which the child can intervene with the proposed actions and the teacher is the one who mediates this process of understanding or internalising knowledge. For Vygotsky (2000), it involves the construction of concepts and does not happen mechanically, as it is not limited to the process of listening and repeating ready-made actions. It is necessary to build new knowledge with the child, taking into account what they already know in order to arrive at the new knowledge. For this to happen, the actions carried out by the teacher must give meaning to the children's actions. For concepts to be formed, it's important for children to be able to solve problems, as this helps to develop their thinking.

> [...] if the environment doesn't present adolescents with any of these tasks, doesn't make new demands on them and doesn't stimulate their intellect by providing them with a series of new objects, their reasoning won't be able to reach the highest stages, or will only reach them with great delay (VYGOTSKY, 2000, p. 73).

The conclusion is that it is necessary to carry out pedagogical work with interaction and mediation in daily practices. The children who had difficulties with colours were able to differentiate between them after using different pedagogical actions. However, the concepts of laterality (right/left) were not understood by all the children, indicating that this concept is more complex to understand for pre-school children. The content worked on needs to be associated with something concrete (a reference).

The next chapter presents the continuity of the process with the children, in relation to the formation of concepts, through individual and collective actions worked on by the researcher and the teacher.

CHAPTER 2

LATERALITY IN SPATIAL ORGANISATION

2.1 The two sides of the body

This chapter presents data researched in 2017, with children in the six (6) age group, enrolled in the first year of the Nereu Ramos Municipal School located in the municipality of Itapejara **D'Oeste-** PR, with the Pre-school class (Pre II E), continuing with the class researched in 2016, which became the 1st year C class. As is typical of every school, there is a turnover of children, which meant that three (3) children who were part of the research went to other schools and four (4) new children joined this class, thus making up the universe of 25 children.

The first action worked on with the children was diagnostic, trying to identify whether the children recognised the colours yellow/green/blue/red as shown in figure 1. The interview showed that all the children recognised the colours, which means they had learned.

The second proposed action (figure 9) proved difficult for more than 40% of the children (data can be seen in Table 5) and showed that there was a need to resume work involving the topological relationships of right/left/front/back.

In order to form the concept of right/left/forward/backward, he relies on Vygotsky (2009), who emphasises that it is necessary to develop certain activities that promote the child's mutual understanding of the proposed content. When the content acquires meaning, the child builds the concept. The table below shows the proportion of children who have identified the process. According to Le Boulch (1982), it is necessary to establish mutual relationships between the organism and the environment so that children are able to form an image of their own body, thus making it the centrepiece of the proposed actions.

Table 5: Children's understanding of right/left in 2017

Topological relationships	Children who had no difficulties	Percentage	Children who had difficulties	Percentage
Behind	15	60%	10	40%
Front	15	60%	10	40%
Right	13	52%	12	48%
Left	12	48%	13	52%

Source: CHIAPETTI (2017).

It was a considerable number of children who needed to be retaught how to develop their laterality. With this data in hand, the teacher worked on the action presented below:

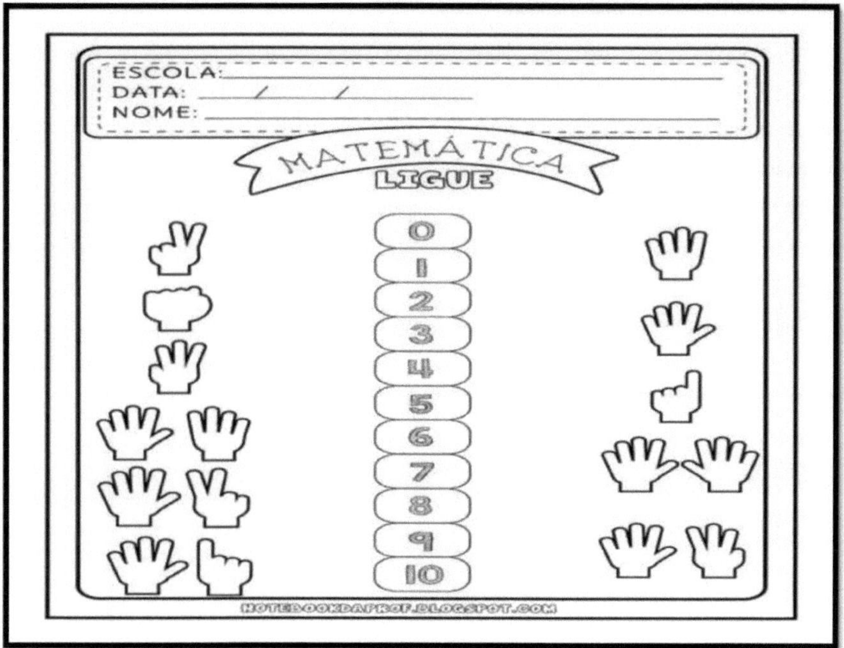

FIGURE 14: The use of right and left when teaching numerals

Fonte: https://br.pinterest.com/pin/748934613003512266/. Accessed on 22 April 2017.

The children related the number of fingers on their right hand to the number of fingers on their left hand, in order to link them to the corresponding numeral. The aim of this activity was to relate maths content to the concepts of right/left topological relationships.

In order for the child to build a concept, they need to socialise their **thinking, so "[...] words, which have not yet reached the level of** fully developed concepts, imitate their function and can serve as a means of **communication and understanding (VIGOTSKI, 2009, p. 159).**

It is through perception, action and thought that the child begins to associate the existing elements in the action proposed by the teacher. This stage is called **the "[...] first stage of concept** formation [...]" (VIGOTSKI, 2009, p. 175).

It involves the process of concept formation and the child builds a syncretic image, a collection of objects. At this stage, the child chooses information at random and discovers the information contained in the actions through error. In the second stage, the child begins to arrange the figures spatially and their perception transforms the heaps of objects present in the first stage into spatial and temporal encounters with a certain meaning. In other words, complexity takes shape for the child. The third stage marks the process of concept formation, where all children attribute the same meaning to objects (VIGOTSKI, 2009). This process occurs when all children construct the concept of each word, universalising knowledge.

Vygotsky (2009) also states that the differences that exist at the moment of learning are established by the links that the child has with the basis of the meaning of the new words (in the case of this research, right/left/front/back) that come to be understood and which are part of a process of stages established between syncretic links (when the child overcomes egocentrism and no longer confuses their impressions with the relationships between the objects under study). In this way, the child moves away from syncretism and towards objective thinking, which is not yet the conceptual thinking present, for example, in an adolescent.

In the following action, we worked on the right hand relationship with the numerals in the left-hand column and with the right-hand column where the numerals are written out in full.

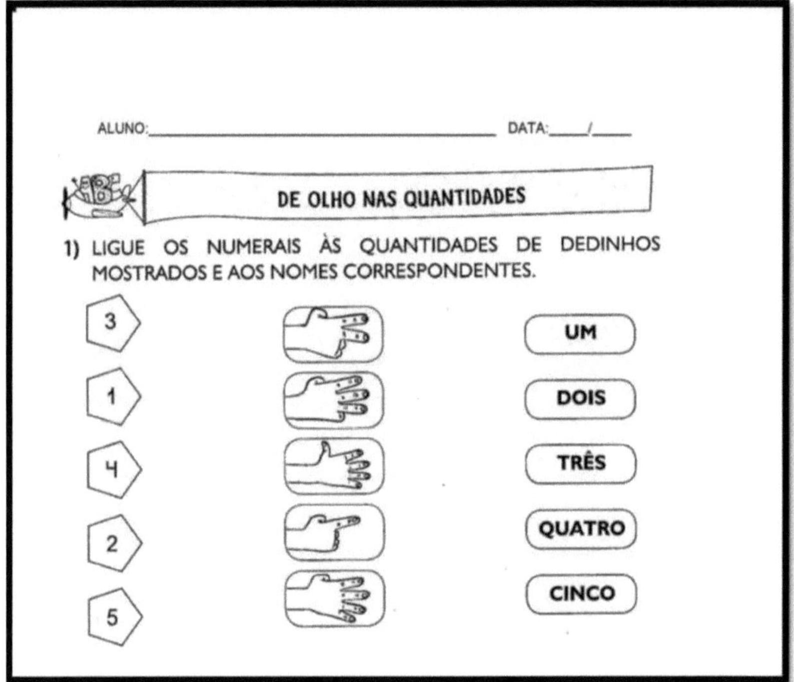

FIGURE 15: Right and left in the study of maths

Source: http://ananhaia.blogspot.com.br/2011/03/numeroquantidade.html. Accessed on 22 April 2017.

Laterality is an important aspect of spatial organisation, as it involves mastering the sides of the body. Almeida and Passini (2002) recommend first starting with actions that involve the body and then getting the child to visualise the information and thus build up the information in their mind. Once the child has formulated the concept, they can transfer it (right/left) to other objects in the classroom. In this way, the child is able to establish links with orientation references, because their

46

perception enables them to

understand the relationship between objects and the concepts of right/left/front/back.

In order to understand the spatial relationships in the body schema, represented in this action, the child needs to build up references to the right and left sides in the representation, as well as the concepts of front and back.The teacher, through this action, worked with the contents of Portuguese.

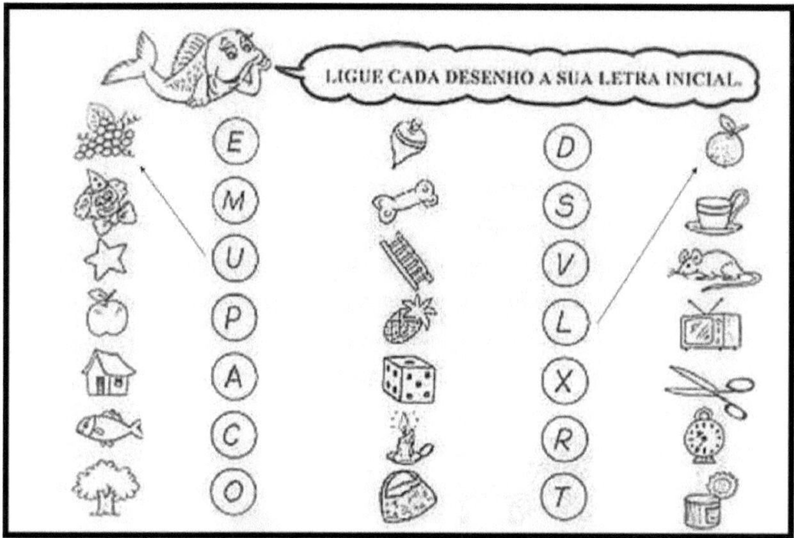

FIGURE 16: Right and left in Portuguese language teaching

Source:http://www.atividadesedesenhos.com/2012/09/exercicios-dia-mundial-da- alfabetizacao_7906.html. Accessed on 22 April 2017.

In this action, the child relates the left-hand column, where the images are, to the initial letter of each word, for example: the image of the grape should be linked to the letter U in the next column, just as the orange in the right-hand column should be linked to the letter L. The teacher took advantage of this action to work on another concept, that of the centre column.

In the second stage of the research, thirteen (13) children still couldn't identify the concepts of right and left. In the first stage, two (2) were unable to do so. On this day, one of the children reported that after being told by the teacher that **she should put herself in the place of the "boy"** in the picture, she **said:** *"Poofesoora if my right hand is the same as the "boy's"* (at this point the child *looked at her hand and the hand of the "boy" in the picture) then my right is the same as his and my left too?"* **(Samoei).**

With this account from the child and based on Vygotsky (2009), when he mentions that children are able to establish links between their initial thinking and the principles present in more complex forms of thought. It can **thus be said that** "[...] in order to understand the other person's discourse [...] we need to **understand their thinking"**. (VIGOTSKI, 2009, p. 481).

After all the children had completed the action, a new activity was presented, in figure 9, which has **the "boy" as its reference. What differentiates** this one from the previous one is that this time, the children had to colour the board with the colour indicated by the teacher, which were the same as those shown in activity 1 (red/green/blue/yellow).

The graphic semiology used in this action can be understood as a set of information that guides the development of activities that make use of symbols or conventions (colours) as information characterisers. Semiotics makes use of different languages and signs to understand an object of investigation, based on visual perception (ARCHELA, 2001).

When the action is drawn up using graphic rules, the information is immediately visualised and the action is no longer a simple illustration. For this reason, the relationship between the data to be obtained from the research and the graphic representation, activity 17, is the starting point for characterising perceptual properties (green/front- yellow/back- red/left- blue/right). Visual variables promote orientation through colour, helping children to develop their perceptual ability through colour association (ARCHELA, 2001).

It was also carried out individually, which showed that some children still had difficulties relating the colour variable to the position requested.

The following activity was used as a tool in the interview with the children so that they could identify the colours, as indicated by the researcher (green/front- yellow/back- red/left- blue/right).

FIGURE 17: Action 17 - Identifying colours and their order

Source: Available at <http://www.todoestudo.com.br/geografia/pontos-cardeais. Accessed on 20 October 2016. Adapted by CHIAPETTI (2017).

The children carried out the action individually. Each child coloured the rectangle **in front of the "boy"** green. Then yellow for the rectangle behind the **"boy". Then** red for the rectangle on the left and finally blue for the rectangle on the right.

According to Archela (2001), associative perception means that the visual variable colour used in the pedagogical action is not confused when it comes to orientation. The aim of this action was to transpose the data present in the oral language (front/back/right/left) when using the visual variables (green/yellow/red/blue).

This action was carried out with the aim of checking whether the children had related the colour variable indicated to its position in the illustration. We realised that some children still had difficulties. For this reason, a teaching sequence was carried out using concrete objects (dolls/cards and the children themselves).

The sequence took place over three days: the class was organised into groups of four children. Each group was given a doll, and it's worth remembering that at this point there was some resistance on the part of the boys to accepting the doll as part of the action presented in the activity. After explaining what the sequence would be like and everyone understanding and being willing to do what was being asked of them, in addition to the doll, each group was given four EVA cards in the colours red/green/blue/yellow, as can be seen in the image below:

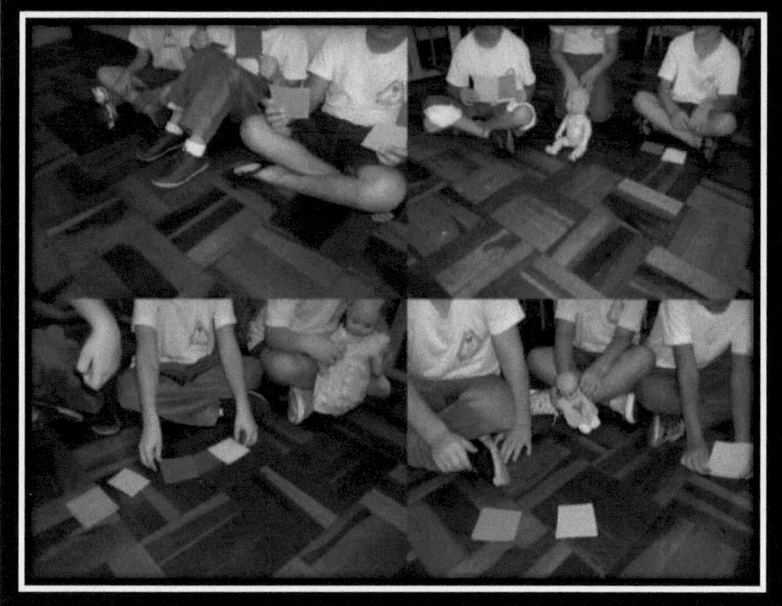

FIGURE 18: Teaching sequence 1, first stage

The groups were organised according to the following criteria: as the boys initially refused to take part in the proposed actions because of the use of the doll, the girls were given the dolls and chose their companions who accepted the nomination in this way. There were a total of six (6) groups of four (4) children and one (1) group had five, due to the odd number of children, 25. Next, one of the children in the group placed the requested card in the position indicated, as can be seen in the second stage (figure 19) of this didactic sequence.

> The construction of the map using the monosemic system requires the correct application of visual variables to each question transcribed visually. Thus, in order to represent information, it is important to carefully observe the significant properties of the visual variables that will be used to transcribe the information from written to graphic language (ARCHELA, 2001, p. 46).

Ideally, when children are presented with an activity, they should feel the need to carry it out. In other words, they need specific motives. These motives are guided by emotions and feelings, which make them develop the action in such a way that the object and objectivity of the activity lead them to an action, which is the internalisation of the knowledge presented by the teacher. This internalisation must promote the transformation of thought through mental acts, mediated by language (LEONTIEV, 1978).

The actions developed by children are conscious processes and delimit the conditions necessary to achieve a final result which is also the result of a socio-historical process, as it involves interaction with peers and the mediation of the teacher (LEONTIEV, 1978).

When children perform collective actions, it can happen that one of them is the protagonist in front of the others. For this reason, the teacher needs to select situations in which they can all act. It is necessary to present actions where progress is made in the cycle, as this favours personal relationships when they develop a joint action (BASSEDAS; HUGUET; SOLÉ, 2007).

In **this sense,** "for these activities to be of interest to children, it is important that the teacher creates a climate in which they are interested and enjoy the activity". (BASSEDAS; HUGUET; SOLÉ, 2007, p. 157).

The image below shows part of the action carried out by the children.

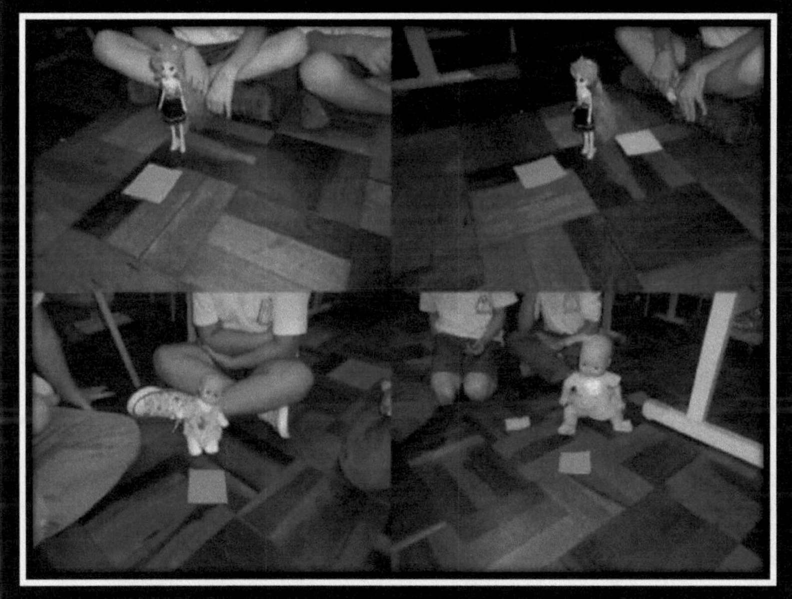

FIGURE 19: Teaching sequence 1, second stage

Source: CHIAPETTI (2017).

Each group placed their doll in the centre and the researcher asked them to place them in the sequence stipulated. Green in front, yellow behind, red on the left and blue on the right. The pedagogical action sought to bring the real closer to the abstract and the result pleased the children who liked the ***"brizaadeira", a* term** used by the children themselves.

In this sense, the language used by children to interact between action and knowledge is also socially produced and loaded with meaning, thus forming consciousness. For Leontiev

(1978), consciousness provides the basis for perception to take place; this perception is constituted through the relationships that are established with colleagues and with the proposed action.

The appropriate use of visual variables allows children to perceive what is being represented, although in order to be understood, graphic semiology requires significant teaching time. Different operations are part of the work of graphic semiology, such as thematic maps that make use of symbols characterising the information (legends).The diagnosis showed that all the children were able to talk to each other and, if one of them didn't get it right, the others promptly corrected them. This attitude promoted a pleasant dialogue and an attempt to get everyone to do what was being proposed correctly. The conclusion of the action can be summarised below:

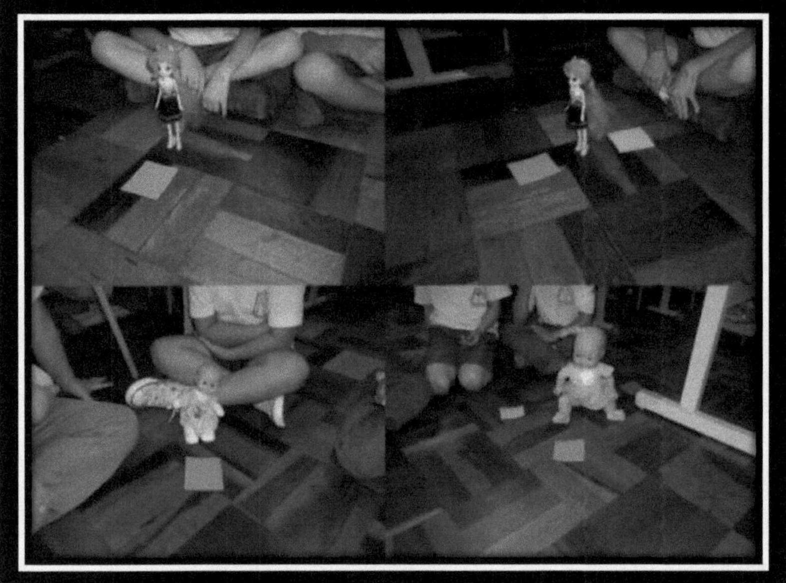

FIGURE 20: Didactic sequence 1, third stage

The same action was carried out, this time with the children in place of the doll and they enjoyed it, as they were asked to each have their cards (green/yellow/blue/red) to carry out the action, which this time had the child's body as the reference point.

> The first condition of all activity is necessity. However, necessity in itself cannot determine the concrete orientation of an activity, because it is only in the object of the activity that it finds its determination: it must, so to speak, find itself in it. Once the need finds its determination in the object (if it is "objective" in it), this object becomes the motive for the activity, the thing that stimulates it (LEONTIEV, 1978, p. 107-108).

The researcher mediated the action so that each colour was placed in the required position. Green was placed at the front, followed by yellow at the back. When it came to red, which was to be placed on the left, many children got confused because they were in opposite positions to the others, so red was changed to the colour blue and, as they are all right-handed in this class, they were given the hint that the right was the hand they were all writing with. In this way, almost all of them were able to relate the colour blue to the right regardless of where they were.

Leontiev's Activity Theory (2005) emphasises that memory, imagination, thought and emotion make up distinct forms of activity, but that thinking and doing need to come together.

One of the boys was confused and then a colleague who was standing next to him got out of her position, raised the boy's right hand and said: "This is your *right* hand, *it's not the same as mine because I'm facing you."* (Dayane)

When social interaction takes place, the child acquires characteristics of the social environment in which he or she is inserted, and learning takes place through the child's interaction with his or her teacher or with other more competent children. The dialogue between the exterior and interior of the individual forms mental actions. For this to happen, exchanges are fundamental, because it is through these (exchanges) that the capacity for internalisation arises, which can manifest itself both intellectually and verbally, but this manifestation requires contact with someone more experienced. Interaction with peers and the teacher brings out the potential of those who are learning, thus activating the procedural schemas present in cognition (VYGOTSKY, 1987).

The development of cognitive structures needs to take into account certain elements such as the child's inter- and intrapersonal level and their internalisation phase, because each person has a different capacity to learn with interaction as a support. Therefore, interaction can be understood as the relationship between what the child can do on their own and what they can do with the help of someone more experienced (VIGOTSKII, 1998).

The teacher who mediates knowledge offers opportunities to raise the level of knowledge and, consequently, learning takes place. Learning cannot only be seen as the result of interaction (VIGOTSKII, 1998).

> [...] the maturation process prepares and makes possible a certain learning process, while the learning process stimulates, as it were, the maturation process and advances it to a certain degree (VYGOTSKY, 2005, p. 4).

For Vygotsky (2004), the formal teaching and learning situations that take place at school are intentional and offer greater possibilities for the development of both children and teachers.

For this reason, the most important factor in the teaching and learning process is the social environment, which at school needs to be systematised, planned, structured and organised so that it promotes relationships of exchange of experiences that maximise the possibilities for the development of all children (VYGOTSKY, 2004).

Children reflect the social environment in which they live and their adaptation occurs through someone who is at a higher level of understanding or knowledge than they are. The image below is a record of the moments of interaction between the children during the actions presented by the researcher, aimed at forming the concepts of front/back/right/left.

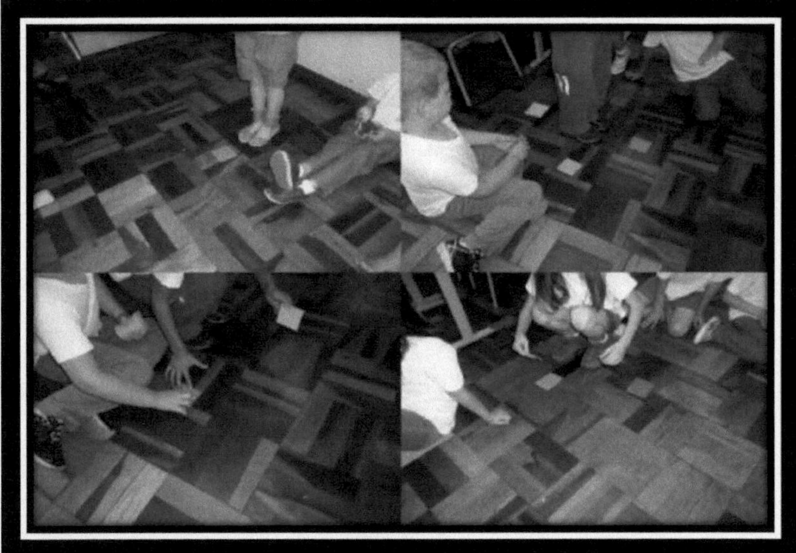

FIGURE 21: Didactic sequence 1, fourth stage
Source: CHIAPETTI (2017).

To the extent that children understand what is being proposed in the action, they begin to enjoy and take an interest in studying it, and it ceases to be an obstacle to their learning (FRANCISCHETT, 1998).

Learning, based on moments of interaction, favours the development of communication and makes it possible to raise the level of cognitive processes: "[...] the entire learning process is a source of development that activates numerous processes that could not develop on their own **without learning"**. (VYGOTSKY, 2005, p. 15).

According to Vygotskyi (1998), learning first takes place at external or social levels, and only then is it appropriated and internalised. Internalisation therefore takes place through communicative mediation between those involved.

Vygotsky (2004) emphasises the importance of previously examining children's knowledge in order to understand what arouses their interest and motivates them, so as to make them active subjects in their learning. By doing this, the teacher is in a better position to find out **"what" and "how" to teach** their children.

Vygotsky (2004) states that the trajectory of cognitive development does not start from the individual to what is socialised, but from the socialised to the individual. Thus, during the process of thought formation, the child triggers actions in which the imaginary represents reality, thus modelling thought. In these actions, external activities are transformed into internal activities.

The apparent homogeneity of children in terms of their level of development shows differences in their possibilities for learning and development. However, a child's Level of Potential Development cannot be understood as a cause or an effect, because imitation, carried out by the child at school age, is a propellant for learning, which occurs when opportunities are provided for social interaction to take place (VYGOTSKY, 1984). When working on the didactic sequence in Geography lessons, the teacher also included actions contained in the maths textbook, as shown in the following example:

FIGURE 22: Front, back, top and bottom

Source: DANTE, 2014, p.16.
The teacher read through the activity with the children. She talked about the concepts above, below, in front and behind. The girls recalled the use of the dolls, while the boys talked about the cards with the colours used (green/yellow/blue/red). The teacher dialogued with the children about questions in the textbook. This activity helped the children relate, through representation, what top,

front, back and bottom mean.

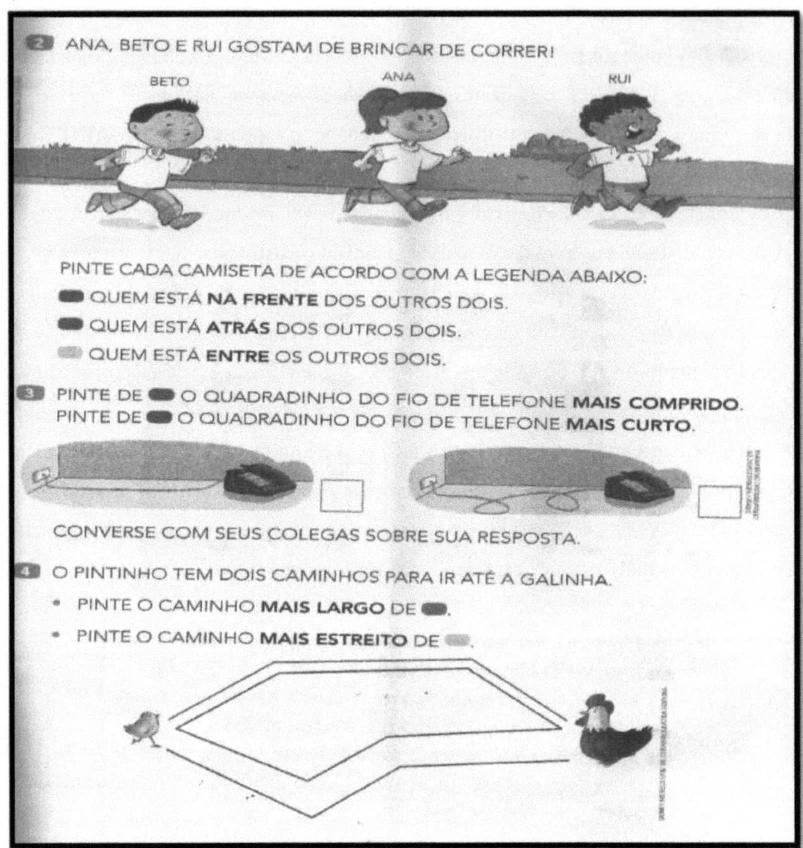

FIGURE 23: Relationship between colours and position

Source: DANTE, 2014, p.17.

The teacher introduced new concepts such as: between, short, long, wide and narrow, as well as in front of and behind. She worked with the position of each child in the classroom and the position of certain objects present, such as cupboards, fans and rubbish bins that don't appear in the picture in the textbook, but which are present in the classroom.The teacher dialogued with the children about the questions and concepts and then the children, in pairs, carried out the activity in the book itself. They made the relationship between the object and its position in the classroom. Below is the image the teacher worked on in the textbook:

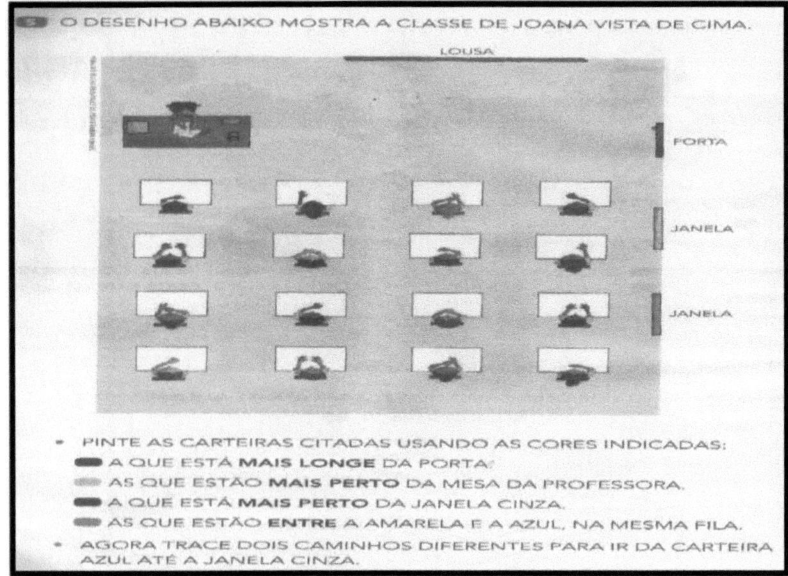

FIGURE 24: Your place in the classroom

Source: DANTE, 2014, p.18.

The teacher added another two (2) concepts: near and far, and the colour variable is something that is very present in the children's literacy process. By making the connection between colour and the proposed action, she enables children to learn in a meaningful way. Another action:

FIGURE 25: The use of topological relationships in children's storytelling

Source: DANTE, 2014, p.19.

The teacher used the previous action to work on orientation, reinforcing the concepts of right and left. She read the statements to the children, who responded as requested. The sequence of this action is shown below:

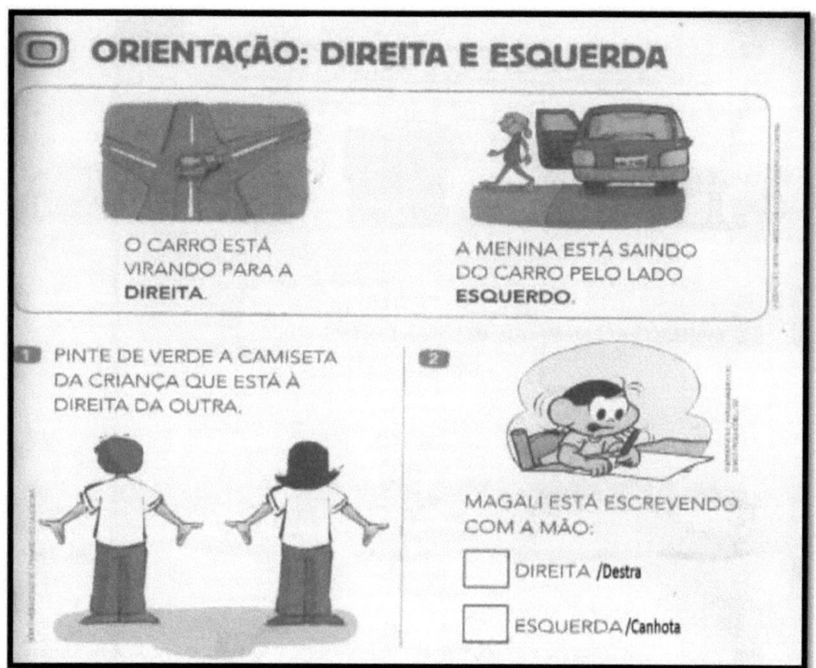

FIGURE 26: Orientation
Source: DANTE, 2014, p. 27, adapted by CHIAPETTI (2017).

Some children managed to realise that they were in the same position as the children in item one (1) of the picture. Other children found it difficult to understand that "Magali" writes left-handed and not right-handed, because she is facing the reader. The teacher had to explain the new fact.

It is worth criticising image 2 of figure 26 presented by the textbook, which, by placing this image at this stage of development, confuses the child and makes learning difficult because it requires a more in-depth understanding of laterality.

This action was also carried out collectively and the teacher explained right/straight and left/tail, followed by other activities, as shown below:

FIGURE 27: The relationship between right and left using animals as a reference point

Source: DANTE, 2014, p. 28.

The children spoke the answers as the teacher read the question, which made it easier for those children who still have difficulty differentiating right from left.

The result showed that all the children were able to identify the colours in Figure 27. Four (4) new concepts related to the colours used since the beginning of the research were inserted, presented in the following action:

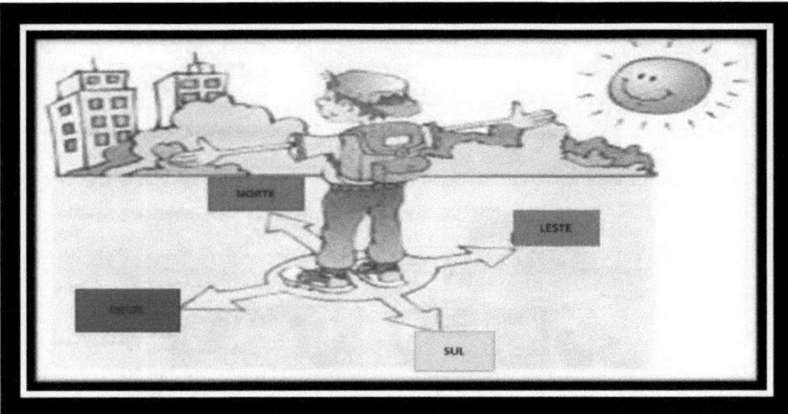

FIGURE 28: Identifying the cardinal points
Source: Available at <http://www.todoestudo.com.br/geografia/pontos-cardeais. Accessed on 20 October 2016. Adapted by CHIAPETTI (2017).

The aim of this action was to incorporate the concepts of North/South/East/West, which are important for teaching Geography as they lead to orientation and localisation. The children found it difficult to relate colour to the new concepts. The children continued to relate the colour to the concepts of front/back/left/right. Three children correctly related green to North, yellow to South, red to West and blue to East, as can be seen in the table below:

Table 6- Cardinal directions

Concepts	Children who had no difficulty with the concepts	Percentage of hits	Percentage related to difficulties
North	8	32%	68%
South	7	28%	72%
East	3	12%	88%
West	9	36%	64%

Organisation: CHIAPETTI (2017).

The difficulty in relating colour to the new concept may be due to the fact that in order to understand it, the child needs to create a mental representation. For this reason, the researcher developed a new teaching sequence the following week. One (1) concept was worked on at a time, and the children were able to transpose the real thing onto paper. Also by visualising the image of the feet represented on cardboard. In this action, each child was able to place their feet on the representation. The action helped to form the concepts. The record can be seen in the image below:

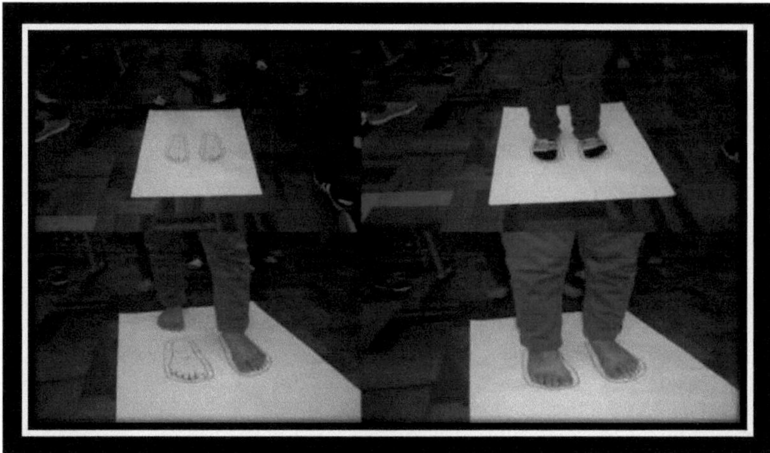

FIGURE 29: Teaching sequence 2, first stage

Each child was asked to place their foot on the drawing that was glued down. This helped them to understand the correct position of the foot, analysing the front of the foot (toes) and the back of the foot (heel), right foot and left foot. This action was important for the child to realise the position of **the "boy" shown in the image** and the relationship with the child's position **in relation to this "boy".** It also helped them understand cartographic representation and introduced them to map reading.

All the children took part and loved putting their feet on the card. The concept of front and back was related to a part of the child's body.

This helped them to construct, through their own experiences, the representation of front/back/right/left.

The children were able to relate the directions to the graphic representation. In the second stage of this didactic sequence, the children were asked where North was. When they identified it, they were given a card with the word North written on it and placed in the position indicated. Then they went to the end of the queue and, once they had all identified it, the concept of South was introduced and the action repeated the same procedure as for the word North. Once identified, the child placed the token in the South and went to the end of the queue, while those who couldn't identify it received help from their classmates. The following image shows the step-by-step process.

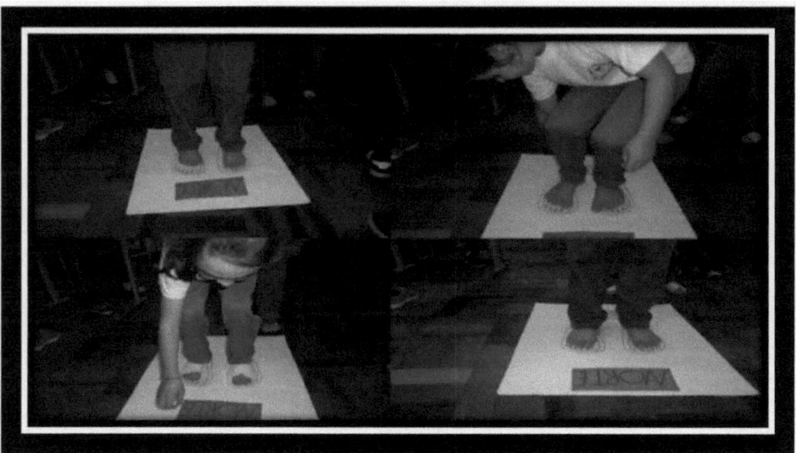

FIGURE 30: Teaching sequence 2, second stage

The children took part in the action and, collectively, those who had difficulty answering were helped by their mates. It's worth noting that so far, none of the children have noticed that the colours of the cards were the same as in the previous actions; they were more concerned with identifying the concepts written on the cards. It is believed that, due to their difficulty in reading, many of them wanted to show that they could read the word. When they had all managed to complete the action, they moved on to the next stage of the didactic sequence: East and West, as can be seen in the picture below.

FIGURE 31: Teaching sequence 2, third stage

Source: CHIAPETTI (2017).

Working in this way, all the children managed to relate the four (4) concepts. Those who had difficulty relating the concept to the position indicated on the card were promptly helped by their classmates.

Each child was asked to work individually on activity 32, which took place as follows: each child was given a sheet of sulphite with two feet drawn on it. These were used as a reference to define the position of the new concepts. The children were very excited to see two small feet printed on the sheet, they all liked it, they thought it was ***"cute"* (an expression** used by the children). After the comments, they were asked to take a green pencil and colour a square in the North position, then yellow and do the same for the South, with red the West and blue the East, as we can see in the image below:

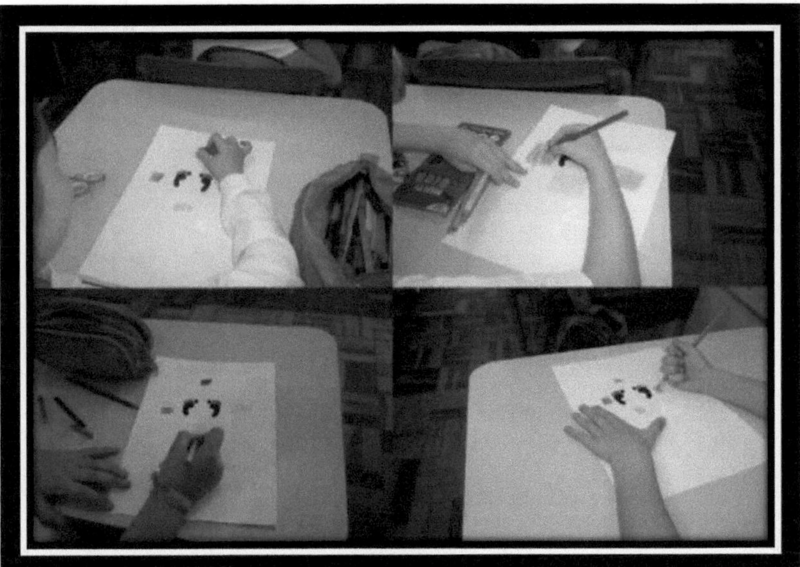

FIGURE 32: Teaching sequence 2, fourth stage

Source: CHIAPETTI (2017).

All of them correctly identified the positions and the corresponding colours. This showed that they were getting closer to the real thing. This is a way of helping children to build important geography concepts in the early years of primary school.

However, in the interview, the children had difficulty identifying the concepts of North/South/East/West. Again, they had difficulty making the relationship between north/front, south/back, east/right and west/left. As the aim was to find out what had gone wrong, we gave the children four (4) EVA cards, in the colours green/yellow/blue/red. Each child then wrote down the four directions: a) Green card - North; b) Yellow card - South; c) Blue card - East; d) Red card - West. This helped the children relate the colours to the concepts, as shown in the picture below:

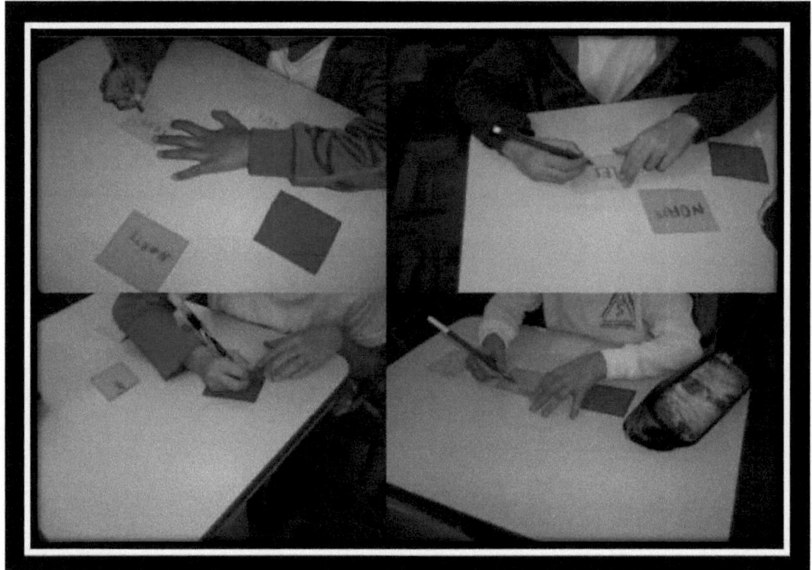

FIGURE 33: Teaching sequence 3, first stage

Source: CHIAPETTI (2017).

After writing the concepts on the cards, each child was asked to hold up the card that represented North. Each child who held up a card of a different colour immediately changed cards to find the right colour while looking at their classmate. This was repeated with the concepts of South/East/West. It should be noted that whenever each child had difficulties, they corrected themselves when they saw that their card was different from the colour of the other children's cards.

Next, each child was asked to place the cards in an orderly fashion on their desk, as if they were represented in the image worked on in activity 28. Thus, the card with the concept North, in the North of the

wallet. They had to place it in front of them, *"away from our bodies",* **as** some children said to those who had difficulty identifying the north of the wallet, even showing it with gestures. When asked to place the yellow card on the south of the desk, the same children who had spoken earlier immediately said: **"Now** *it's near the barrign, tied to* **my foot".** When asked to place the blue card on the east side of the desk, some of the children had difficulty defining which side it was. When they realised their uncertainty, one of the pupils said: "It's **the** *side of the hand you write* on (Khiara). When they were asked to place the red card on the west, they all identified it.

FIGURE 34: Teaching sequence 3, second stage

The action was repeated a few times and each child was asked to come forward (one at a time) and place the cards (this time larger and made by the researcher) as recommended. They were given the green card with the word North written on it and asked to place it where they thought North was, the yellow card where they thought South was, the blue one in the East and the red one in the West. In this action, two children still confused the positions and, as agreed with the class before the action began,

None of them could say where the mistake was. Only after the action had been completed would the class be asked if the positions were correct; when they weren't, the child who had got the positions mixed up would ask for help from a classmate they had chosen to help them position the tokens correctly.

According to Vygotsky (2000), knowledge is procedural and the teacher has the role of mediating this process, not facilitating it. In this way, "[...] the child is able to do more with the help of another person **(teachers or peers) than they would on their own [...]**" (CAVALCANTI, 2015, p. 194). This scene can be seen in the first image in figure 35, where the child has placed the card with the inscription South where it would be East, East where it would be South and North where it would be South.

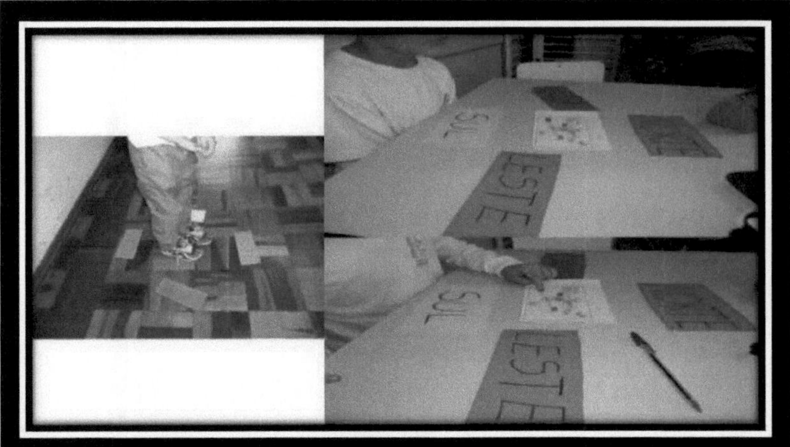

FIGURE 35: Teaching sequence 3, third stage

Source: CHIAPETTI (2017).

Once this collective stage had been completed, individual work began, with most of the children managing to position the cards correctly. However, two children still had difficulties. Each child was positioned as they were in the room, so they were able to identify East and West. However, they still had difficulty with North and South. So they were given the cards and asked to place them as they were written, the card with the word North on it on North, the card with the word South on it on South,

the card with the word East on it and the card with the word West on it, so the two children managed to position it correctly.

The card was then placed in the middle of the activity and they were asked to analyse the similarity in the colours, positions and writing of the words (these two children have no reading skills). After analysing the cards, one of the children said: "The *colours are the same and it looks like the words are too"* (Braian), this scene is present in the last two images above. The child pointed with his finger to everything that was asked for correctly. The other child also placed the cards correctly in the positions requested, concluding our research at this school.

In the early years, children need to constantly relate, visualise and handle information in order to form concepts, because "[...] the development of a more comprehensive and abstract way of thinking about geography therefore requires the formation of concepts". (CAVALCANTI, 2015, p. 201).

The notion of North/South/East/West through the topological relationships front/back/right/left can **contribute "[...] to the development of** the skills necessary for the child's movements, [...]" (CAVALCANTI, 2015, p. 203). In this sense, children aged five (5) and six (6) have the topological relationships front/back/right/left as the basis for forming important concepts for Geography, as it is the structural basis of the content present in the school curriculum that works with the organisation of the space lived by the child, which is their street, their neighbourhood, their school and their city.

The topological relationships front/back/right/left are the first relationships that children establish with their bodies and this will direct their actions, given that during this period they display a certain egocentrism and project what they observe from their point of view (CALLAI, 2005).

The following chapter will present the research carried out in three (3) other schools in the municipality of Itapejara **D'Oeste:** Escola Municipal Prof. Pedro Viriato Parigot de Souza, Escola Municipal Josafat Kmita and Escola Municipal do Campo Valentim Biazussi, the latter of which is multi-serialised and involves pre-school (Pre I and II), first and second year children. The research was extended to the other schools to see if the use of certain pedagogical actions can help in the process of concept formation, giving greater credibility to this dissertation, since the larger number of subjects participating in this research presents information that is closer to the real thing.

CHAPTER 3

CONCEPT FORMATION

3.1 The construction of concepts in Geography

The process of concept formation with children aged five (5) and six (6) at the Nereu Ramos Municipal School was expanded to the Irmão Josafat Kmita Municipal School, the Prof. Pedro Viriato Parigot de Souza Municipal School and the Valentim Biazussi Municipal School in the countryside because the need arose to test the actions in other environments in order to better substantiate the research data.

The period between the interviews in the three (3) schools was five weeks. Only the interview involving action 17 - Identifying colours and their order had to be repeated, and only at the Prof. Pedro Viriato Parigot de Souza Municipal School, where five (5) children had difficulty.

The different time frame used by the research when working on the proposed actions between the schools can bring about significant changes in the process of understanding concepts (north/front, south/back, east/right and west/left) and colours (green/blue/red/yellow), as children's cognitive development interferes with their ability to assimilate.

Vygotsky (1984) emphasises that children project themselves onto the activities that adults present to them and begin to incorporate them into their way of acting and thinking. This makes them change and evolve.

Pedagogical actions help the child's intellectual development process because they trigger actions. The objects they handle complement their geography repertoire (VYGOTSKY, 1984).

The actions represent different ways of reconciling the graphic with the real, contributing to the modelling of the internal processes that occur in thought (VYGOTSKY, 1984). The following are the actions that formed part of the investigative script for this research.

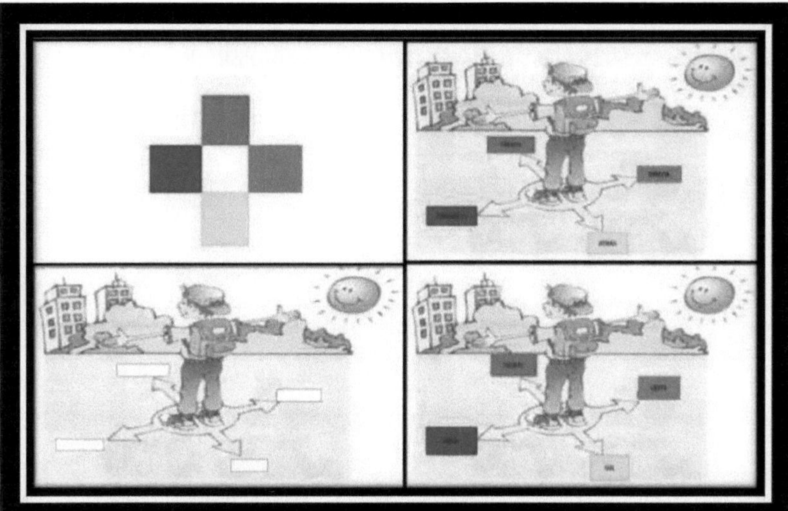

Figure 36: Pedagogical actions/investigative itinerary
Source: CHIAPETTI (2017).

In this colour identification action, the diagnosis showed that none of the 63 (sixty-three) children had any difficulty identifying the colours green/blue/red/yellow.

In a conversation with the teacher at the Valentim Biazussi Municipal School in the countryside, she reported that she had worked on a pedagogical activity in which the children were placed in a circle and given two balloons with the colours blue and red, then the children had to pass the blue balloon to the person on their right and the red balloon to the person on their left.

Next, the children were placed in a line behind each other and the first child in the line was given a yellow balloon which, on the teacher's signal, they were to hand to whoever was behind them, then the last child was given a green balloon, which they were to hand to the child in front of them. According to the teacher, this action helped the children to identify the colours and the order proposed in the research. According to the teacher, this helped the younger children in particular, as the school works with small classes.

classrooms. Children aged four (4) to seven (7) are in the same classroom.

The other teachers reported that they hadn't worked on any specific action for the interview, because, according to them, colours and order are worked on almost daily in all subjects.

As none of the children had any difficulty identifying the colours shown (green/red/blue/yellow), the research was followed up with work on identifying the topological relationships of right/left and front/back. The children from the three (3) schools were shown the same teaching sequence presented at the Nereu Ramos Municipal School. Figure 37 shows this teaching sequence.

FIGURE 37: North/front, south/back, east/right and west/left relationships

Source: CHIAPETTI (2017).

The children were given four (4) EVA cards in the colours green/red/blue/yellow. When asked if they remembered the colours, they all said yes. To check, they were asked to hold up the green card, then the yellow card and so on with the red and blue colours. Once they had realised that the cards had been held up

correctly, they were asked to place the green card at the front of the desk, the yellow card at the back, the blue card on the right-hand side and the red card on the left-hand side of the desk. All the children distributed the cards correctly. As none of the children had any difficulty in carrying out this sequence, the smaller cards were replaced by larger ones where the concepts of North/South/East/West were presented and they were asked to relate the colour to the concept. Some children read what was written, while others managed to relate the concept to the colour. Example: North was in front, and the colour in front was green, so North was green and so on with the other colours.

None of the children had any difficulty identifying the right, left, front and back sides of the representation. However, on this day, the teacher at the Prof Pedro Viriato Parigot de Souza Municipal School told us that one (1) of her children had been diagnosed with Global Developmental Disorder (GDD) which, according to Rotta, Bridi Filho and Bridi (2016), is characterised by a certain degree of impairment in learning and social interaction. Most children with some kind of Global Developmental Disorder need to be monitored during their learning. Even with this medical report, the child showed no difficulties and correctly identified the topological relationships.

Straforini (2008) emphasises that actions that include orality are important for the process of building knowledge, because when children talk to their peers, they receive and transmit information.

> [...] the process of developing concepts themselves does not take place from the inside -intrapersonal- to the outside -interpersonal-, but rather in the opposite direction, i.e. from the outside to the inside, or from the interpersonal to the intrapersonal. The Higher Psychological Processes (HPS) originate in social life, that is, in the subject's participation in activities shared with others (STRAFORINI, 2008, p. 107).

However, we realised that some children still turned to their classmates when they needed to identify the right and left sides. For this reason, we worked on an activity in which they could transpose the real thing onto the graph, as seen in figure 38:

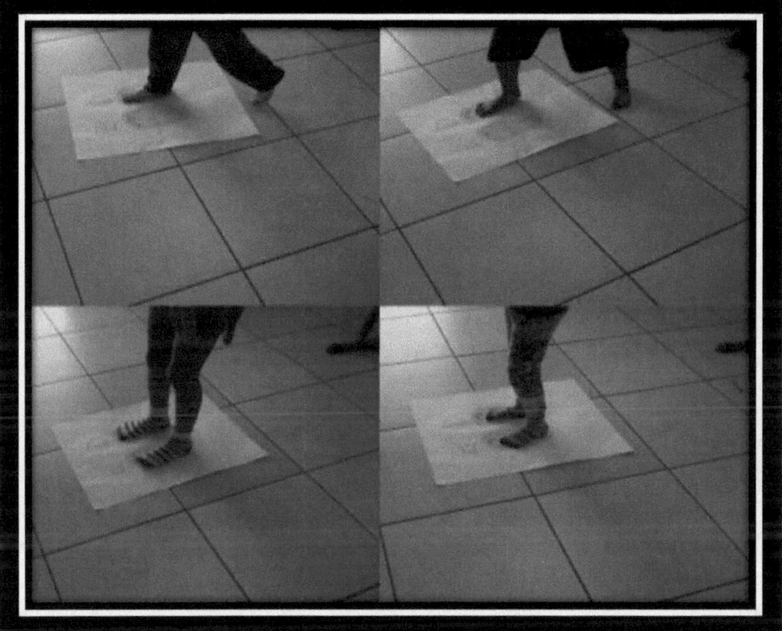

FIGURE 38: Transposition activity from the real to the graph
Source: **CHIAPETTI (2017).**

The children were asked to place their feet on the representation on the poster. They did this individually as requested: first they put their right foot on the representation of their right foot, then their left foot on the representation of their left foot.

For children to understand the concepts used in Geography (right/left), they need to start with direct experiences, with actions that help and facilitate understanding. The proposal shown in figure 38 is an alternative methodology that helps to make children feel like participants in the process of building knowledge. The aim of the proposal is to understand the dynamics and difficulties presented by children during **"learning" in order to present some** possibilities for teaching Geography (CAVALCANTI, 1998).

In everyday life, children encounter different geographical elements. However, their understanding is limited because children are aware of the existence of these elements, but are unaware of how they can use them in different situations **(STRAFORINI, 2008)**. **"Learning to direct one's own** mental processes with the help of words or signs is an integral part of the process of concept formation". (VYGOTSKY, 1989, p. 51) When they managed to relate right/left, the following action was presented:

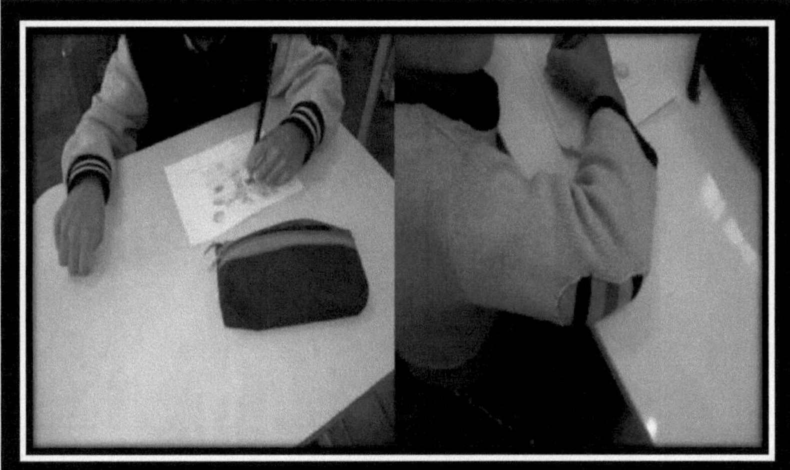

FIGURE 39: Concepts of North, South, East and West

Source: CHIAPETTI (2017).

Whenever the child was asked to identify one of the concepts, they would point to it with their pencil in the colour that represented the concept. For example: when they said South, the child would point with their pencil to yellow, East to blue, West to red and North to green. The children who could read reported that all they had to do was read what was written. Those who had not yet mastered reading, according to what they said, remembered the colours and related them to the concept. The aim was to see how representations can indicate ways of teaching geography.

The children found it easy to carry out the activity. The actions proposed in the four (4) schools in the Itapejara **D'Oeste** municipal education network **helped to affirm that the** children **developed** spatial organisation.

When children are able to locate objects in relation to themselves, they are able to understand directions. Therefore, orientation is an evolutionary process in which pedagogical actions need to develop notions of interiority/exteriority/ intersection/continuity (SANTOS, 2013).

The direction to the right, presented as a reference to **the** sun, is also **marked by the "boy's" open arms, which consolidate** the body's position in **relation to the sun**. Consolidating a system of communication between the representation and the subject. In other words, the interlocutor, giving it living and dynamic meaning by participating in the actions proposed by this research (FRANCISCHETT, 2012).

The symbology used in representations **"are constructions that** replace things and result from an activity (FRANCISCHETT, 2012, p. 103). Representation has become one of the stages in the process of producing knowledge. This research cannot highlight the extent to which the child has actually been informed, but it does organise some actions that offer possibilities for working with geography in the early years of primary school.

According to Straforini (2008), children need to be shown actions that intertwine everyday life with specific knowledge, so that they are faced with situations where they can understand reality, arousing a greater interest in trying to understand.

A child's mental functions develop during interactions between subjects (teachers/students, students/students) who are in constant activity with the world around them (CAVALCANTI, 1998).

Based on Vygotsky, Cavalcanti (1998) emphasises that in order to form a concept, children need to unite and separate elements (synthesis and analysis). This process can be affected by different external and internal conditions, but is generally unique to each child.

The pedagogical actions presented to children need concrete experiences so that they go beyond their limits. However, teaching "[...] Geography in the early grades does not end with its concepts and notions. On the contrary, it starts from them towards a greater vision [...]" (STRAFORINI, 2008, p. 167).

We conclude that the visible difference in the performance of the children surveyed in 2017 is due to the fact that the work was first developed with children who had not yet entered the world of reading and literacy itself, as they were of pre-school age, a period of transition and adaptation between nursery school and primary school.

Those who were in the first year of the Nereu Ramos Municipal School were also interviewed in the first half of 2017, when the literacy process began, which is why they had difficulties, especially with regard to the correlation between laterality and concepts, while those who belonged to the other three schools had already been brought into contact with these concepts by their teachers and the teachers responsible for working with Geography as a diversified curricular component.

Over time, children's knowledge (ability to interpret phenomena) increases, as do their skills (DUARTE, 2002). By the end of the second half of the year, children are better able to learn because their comprehension skills are already sharper, which contributes positively to the process of concept formation.

CHAPTER 4

FINAL CONSIDERATIONS

The pedagogical actions presented in this research show how important it is to teach children and that learning depends a lot on how the content is related to what is visible, to what is known by the child. If the child has difficulty forming concepts, it's because the process is still underway.

The transformations that take place in children's thinking depend a lot on their school environment, which allows them to understand the various aspects involved in making decisions in the face of situations and challenges that require the intervention of a mediator, who is the teacher. However, when interacting with peers and information, peers also become mediators in the process of building knowledge.

Pedagogical actions involve their own language that interweaves knowledge and depends on interpreting information from the symbolic representative level.

It was difficult for some of the pre-school children to recognise the colours green/blue/yellow/red used in this research, even though they had already been to school, as most of them had been at nursery since they were two (2) years old. However, the individual work carried out in the first phase of this research was an opportunity for the teacher, researcher and coordinator to investigate possible flaws in the teaching and learning process. The same happened with the activities involving notions of laterality (right/left/front/back). The colours therefore served as a support (overlapping information) for understanding the topological relationships of back/front/right/left so that the children could relate them to the concepts of the cardinal directions of North/South/East/West.

According to the actions worked on and evaluated, the results show that some proposals are satisfactory, while others are not. Although the content of Geography in the early years is fundamental, as it aims to form important concepts for children to understand their geographical surroundings, in some situations it is not given its due importance. Many of the actions did not provide for debate among the children, they simply followed the stages.

That's why the teacher's role as mediator during the teaching and learning process is so important, so that his or her actions don't become mere tasks.

However, many of the activities developed, especially the didactic sequences and the dead/alive game, are actions that have had a positive outcome, because with these actions the children have been able to name and signify the concepts.

The concepts worked on in this research are important because they help children understand cartography. This research does not detract from lectures or working with textbooks, but it does emphasise the importance of the child's active participation during the process of concept formation.

Children need to be offered didactic support so that they understand that the concepts they are working on are part of their daily lives and help them to perceive the geographical space around them.

Teachers in the early years find it difficult to work with activities that involve the formation of geographical concepts, and this ends up reflecting on the process of teaching content at this stage.

This research became a space for discoveries about methodologies that, for teachers, because they are so simple, end up not being taught. One of the most striking moments was discovering the difficulty that some pre-school children have with colours, something so simple that it sometimes goes unnoticed by the teacher who doesn't take individual action with regard to colours. The research teaches us the importance of individuality when investigating a child's difficulties.

Some of the actions presented give children the opportunity to participate in the construction of knowledge, showing the importance of starting with practical activities and then presenting representations.

Children need to have knowledge associated with something that is part of their daily lives, especially when it comes to laterality. Recognising right and left was the activity the children found most difficult.

North/South/East/West are also known as geographic references or directions and are part of the content of the Geography curriculum in the early grades of primary school. They are extremely important for the formation of knowledge involving location and orientation. When children don't understand the relationship between these concepts, it becomes difficult for them to orientate or localise themselves, or even to understand what these concepts are for.

CHAPTER 5

REFERENCES

ALDEROQUI, Silvia. **Teaching how to think about the city**. In: ALDEROQUI, Silvia; Ppenchansky, Ponpi. Ciudad y ciudadanos: aportes para la ensenanza del mundo urbano. Buenos Aires: Paidós, 2002.

ALMEIDA, Rosângela Doin de; PASSINI, Elza Yasuko. **Geographical space:** teaching and representation. 12 ed., São Paulo: Contexto, 2002.

ALVES, Maria de Fátima. **From repetition to learning:** cognitive development through interaction. Veredas on line - Ensino - 2/2007, P. 41-57 - PPG Linguística/UFJF - Juiz de Fora. Available at < http://www.ufjf.br/revistaveredas/files/2009/12/artigo031.pdf>. Accessed on 26 December 2016.

ARCHELA, Rosely Sampaio. **Contributions of Graphic Semiology to Brazilian Cartography**. Geografia, Londrina, v. 10, n. 1, p. 45-50, jan./jun. 2001.

BASSEDAS, Eulália; HUGUET, Teresa; SOLÉ, Isabel. **Learning and teaching in early childhood education**. Translation by Cristina Maria de Oliveira. Porto Alegre: Artmed, 2007.

BERTIN, Jacques. **The neographic**. Translated by Jayme Antonio Cardoso (UFPR), July/2000. This refers to the work by BERTIN, Jacques. Neography and the Graphic Treatment of Information. Curitiba, Editora da UFPR, 1986.

BRAZIL. **CNE/CEB. Resolution no. 7**, of 14 December 2010. Ministry of Education/National Education Council/Basic Education Chamber. National Curriculum Guidelines for nine-year primary education. Available at <http://portal.mec.gov.br/dmdocuments/rceb007_10.pdf.>. Accessed on 16 July 2017.

CALLAI, Helena Copetti. **Learning to read the world:** geography in the early years of primary school. Caderno cedes, vol. 25, n. 66, p. 227- 247, may/august. Campinas: Unicamp, 2005.

______. **The teaching and research of Geography for the initial years of Primary Education**. Brazilian Journal of Geography Education. Campinas, v. 6, n. 11, p. 06-20, jan./jun.; 2016.

CALLAI, Helena Copetti; CALLAI, Jaeme luiz. **Group, space and time in the early grades.** Boletim gaúcho de Geografia. V. 21, n. 1. Porto Alegre, 1996. Available at http://seer.ufrgs.br/index.php/bgg/article/view/38636/26360. Accessed on 30 January 2017.

CAVALCANTI, Lana de Souza. **Everyday life, pedagogical mediation and concept formation:** Vygotsky's contribution to geography teaching. Caderno cedes, vol. 25, n. 66, p. 185-207, May/August 2005. Available at http://www.cedes.unicamp.br. Accessed on 20/04/2015.

______. **Geography teaching and diversity:** construction of school geographic knowledge and attributions of meaning by the various subjects in the teaching process. In: CASTELLAR, Sônia (org). Geographical education: theories and teaching practices. São Paulo: Contexto, 2007.

______. **Geografia, Escola E Construção de Conhecimentos.**10 ed., Campinas: Papirus, 1998.

DANTE, Luiz Roberto. **Ápis:** mathematical literacy. 2 ed., São Paulo: Ática, 2014.

DUARTE, José B. **Case studies in education. In-depth research with reduced resources and another way of generalising.** Revista Lusófona de Educação, 2008.

DUARTE, Newton. **Activity theory as an approach to educational research.** Perspectiva, Florianópolis. V. 20. N. 02, Jul/Dec. 2002.

______Newton. **Formation of the individual, consciousness and alienation: the human being in the psychology of A. N. Leontiev.** Cadernos CEDES, Campinas, v. 24, n. 62, Apr/2004.

ESMERALDO, Valéria. **Playing and learning with Monica's gang.** São Paulo: Rideel, 2010.

FRANCISCHETT, Mafalda Nesi. **Image-based learning in geography teaching.** In: ANDRES, Juliano; FRANCISCHETT, Mafalda Nesi; AGUIAR, Waldiney Gomes (eds). Teaching Geography: approaches to geocartographic representations and teacher training. Cascavel: EDUNIOESTE, 2012.

______. **The concept of orientation and localisation in textbook images.** ENPEG: 10th National Meeting on Geography Teaching Practice. Porto Alegre. From 30th August to 2nd September 2009.

______. **Building methodological links in cartographic language.** Revista brasileira de Cartografia, Rio de Janeiro, n. 63/4, p. 843-859, Jul/Aug 2014.

______. **Localisation using maps.** Faz Ciência. Francisco Beltrão. v. 2. n. 01. p. 9-22. 1998.

GIORDANI, Ana Claudia Carvalho. **Connecting geography in cyberculture**: international bank of educational objects and digital school. In: CALLAI, Helena Copetti; TOSO, Cláudia Eliane Ilgenfritz. Dialogue
with teachers: citizenship and educational practices. Ijuí: Unijuí, 2015.

GRYMUZA, Alissá Mariane Garcia; RÊGO Rogéria Gaudencio. **Activity theory:** a possibility in maths teaching. Revista Temas em Educação. v. 23, n. 2. Paraíba: UFPB, 2014.

GUERRERO, Ana Lúcia de Araújo. **Contributions of activity theory to the continuing education of geography teachers.** In: CASTELLAR, Sônia (org). Educação geográfica: teorias e práticas docentes. 2 ed., São Paulo: Contexto, 2007.

HENNIG, Georg J. **Metodologia do Ensino de Ciências.** New Perspectives Series. 3 ed., IVIC, Ivan; COELHO, Edgar Pereira. Recife: Massangana, 2010.

IVIC, Ivan; COELHO, Edgar Pereira. **Lev Semionovich Vygotsky.** Recife: Massangana, 2010.

JODELET, D. **Les Représentations sociales**: un domaine en expantion. In: Social Representations. D. Jodelet (ed.). Paris: Press Universitary de France, 1989.

LE BOULCH, Jean. **Psychomotor development:** from birth to six years. Translated by Ana Guardiola Brizolara. Porto Alegre, Artes médicas, 1982.

LEONTIEV, Alexei Nicolaevich. **The development of the psyche.** Translated by Manuel Dias Duarte. Lisbon: Horizonte Universitário, 1978.

______. **The principles of mental development and the problem of mental retardation.** In: Psychology and pedagogy: Psychological bases of learning and development. Translated by Rubens Eduardo Frias. São Paulo: Centauro, 2005.

MARTINS, Onilza Borges; MOSER, Alvino. The concept of mediation in Vygotsky, Leontiev and Wertsch. **Revista Inter saberes**. vol. 7. n.13, p. 8 - 28, jan./jun. 2012.

MONTEIRO, Darlisângela Maria; GHEDIN, Evandro; KRUGER, Marcos Frederico. **Leontiev's epistemology, the relationship between the development of the psyche, culture and its implications for science teaching**. Available at http://www.nutes.ufrj.br/abrapec/vienpec/CR2/p756.pdf. Accessed on 03 Jan 2017.

MORETTI, Vanessa Dias; ASBAHR, Flávia da Silva Ferreira; RIGON, Algacir José. **The human in man: the** theoretical and methodological presuppositions of cultural-historical theory. Psychology & Society. V. 23. N. 3. Minas Gerais: Brazilian Association of Social Psychology, Sep/Dec 2011.

MOURA, Manoel Oriosvaldo de (org). **Pedagogical activity in historical-cultural theory.** Brasília: Liber, 2010.

OLIVEIRA, Marta Kohl de. **Thinking about education: Vygotsky's contributions.** In: Piaget Vygotsky: new contributions to the debate. São Paulo: Ática, 1988.

PRESTES, Zoia Ribeiro. **When it's not the same: Analysing translations of Lev Semionovictch Vigotski in Brazil, repercussions in the educational field**. Brasília: University of Brasília, 2010.

REGO, Teresa Cristina. **Vygotsky:** uma perspectiva histórica-cultural da educação. 22 ed., Petrópolis: Vozes, 2011.

RODRIGUES, Orlando. **Activity theory and transformation through action**. Available at: http://www.administradores.com.br/artigos/negocios/a-teoria-da- atividade-e-a-transformacao-pela-acao/12668/. 19 September 2006. Accessed on 13 March 2016.

ROSA, Paulo Ricardo da Silva. **Vygotsky's theory**. Chapter 5. UFMS Physics Department. Available at: http://pt.slideshare.net/cdim4k/teoria-de-vygotsky. Accessed on 13 March 2016.

ROTTA, Newra Tellechea; BRIDI FILHO, César Augusto; BRIDI, Fabiane Romano de Souza. **Neurology and learning: a** multidisciplinary approach. Porto Alegre: Artmed, 2016.

SÁNCHES VÁSQUEZ, A. **Filosofia da práxis**. 2 ed., Rio de Janeiro: Paz e Terra, 1977.

SANTOS, Clézio. **Cartographic knowledge.** Nova Iguaçu: Agbook, 2013.

SIMIELLI, M. E. R., **Cartography in primary and secondary education**. In: CARLOS, A. F. A. (org). Geography in the classroom. 8 ed., São Paulo: Contexto, 2007.

SIRGADO, Angel Pino. **The concept of semiotic mediation in Vygotsky and its role in explaining the human psyche**. Cadernos Cedes, n. 24. Pensamento e linguagem- Estudos na perspectiva da psicologia soviética. 3 ed., Jul/ 2000.

STRAFORINI, Rafael. **Teaching Geography:** the challenge of the world in the early grades. 2 ed., São Paulo: Annablume, 2008.

TOSO, Cláudia Eliane Ilgenfritz. **Children, space, time and the construction of knowledge.** In: _______________________ . CALLAI, Helena Copetti; TOSO, Cláudia Eliane Ilgenfritz. Dialogue with teachers: citizenship and educational practices. Ijuí: Unijuí, 2015.

VASCONCELLOS, Celso dos Santos. **Constructing knowledge in the classroom.** 17 ed., São

Paulo: Libertad, 2005.

VIGOTSKI, Lev S. **The construction of thought and language**. 2 ed. Paulo Bezerra. São Paulo: WMF Martins Fontes, 2009.

VIGOTSKI, L. S. **Thought and word**. In L. S. Vygotsky. The Construction of Thought and Language. São Paulo: Martins Fontes, 2001.

VYGOTSKY, Lev Semenovich. **Thought and language**. Psychology and Pedagogy Collection. São Paulo: WMF Martins Fontes, 1987.

VIGOTSKII, Lev Semenovich; LURIA; Alexander Romanovich; LEONTIEV, Alex N. **Language, Development and Learning.** Translation by Maria da Pena Villalobos. 11 ed. São Paulo: icon, 2010.

______. **The social formation of the mind**: the development of higher psychological processes. Psychology and Pedagogy Collection. 3 ed., São Paulo: WMF Martins Fontes, 1989.

______. **The Social Formation of the Mind**. 6 ed. José Cipolla Neto, Luis S. M. Barreto and Solange C. Afeche. São Paulo: WMF Martins Fontes, 1998.

VYGOTSKY, L. S. **The social formation of the mind**. São Paulo: Martins Fontes, 1984.

______. **An experimental study of concept formation**. Psychology and Pedagogy Collection. São Paulo: Martins Fontes, 2000.

______. **Pedagogical Psychology**. Translation by Paulo Bezerra. 2 ed., São Paulo: Martins Fontes, 2004.

______. **Learning and Intellectual Development at School Age**. In: Luria, Leontiev, Vygotsky et al. Psychology and Pedagogy. Psychological Bases of Learning and Development. São Paulo: Centauro, 2005.

Yin, Robert K. **Case study**: planning and methods. Translation by Daniel Grassi. 2.ed., Porto Alegre: Bookman, 2001.

ANNEX I

INFORMED CONSENT FORM

"Topological relationships with primary school children in Geography teaching"

Dear Sir or Madam:

We are inviting the child, under your responsibility, to take part in the research **"Topological relations with children in the early years in the teaching of Geography",** to be carried out at the **"school".** The aim of the research is to **"analyse how topological relationships enable children aged 5 to 6 to perceive and understand geographical space".** The child's participation is very

82

important and will be through an **interview with questions aimed at the proposed objective.**

We would like to clarify that the child's participation is completely voluntary, and that you may ask the child to refuse or withdraw from participation at any time, without this entailing any burden or harm to the child. We would also like to clarify that the information will only be used for academic scientific purposes and in such a way as to preserve the child's identity. The names used in the transcriptions will be fictitious.

We would also like to make it clear that neither you nor the child you are responsible for will pay or be paid for taking part. We would like to inform you that this research takes into account and respects the rights laid down in the Statute of the Child and Adolescent (ECA), Federal Law 8069 of 13 July 1990, namely: life, health, food, education, sport, leisure, professionalisation, culture, dignity, respect, freedom and family and community life. We also guarantee that Article 18 of the ECA will be complied with: "It is everyone's duty to watch over the dignity of children and adolescents, keeping them safe from any inhuman, violent, terrifying, vexatious or embarrassing treatment. "

If you have any questions or need further clarification, please contact us **(Dulcinéia Cristina Chiapetti, Rua Sete de Setembro/650 Barra Grande, telephone 999157337 or 35266061, email: dulce_chiapetti@outlook.com).**

This form must be completed in two copies of equal content, one of which must be duly completed, signed and delivered to you.

Itapejara **D'Oeste, ___ 2017.**

Dulcinea Cristina Chiapetti

ID: 5497232/6.

<table>
<tr><td>

_______________________________________ **(NAME OF PERSON RESPONSIBLE FOR PARTICIPANT IN RESEARCH**), having been duly informed about the research procedures, agree to the **voluntary** participation of the child under my responsibility in the research described above.

Signature: _______________________________

Date: ___________________

</td></tr>
</table>

Consent form

STATE UNIVERSITY OF WESTERN PARANÁ- UNIOESTE
POSTGRADUATE PROGRAMME - STRICTO SENSU MASTER'S DEGREE IN GEOGRAPHY

<u>Consent form for children and adolescents (over 6 and under 18)</u>

You are being invited **to take part in the research project "Topological relationships with primary school children in Geography teaching".** Your parents have authorised your participation. We are going to analyse how topological relationships can help in understanding geographical space.

You don't have to take part in the survey if you don't want to. The research will be carried out at the school. The information as well as the results of the research will be used exclusively for scientific-academic purposes.

===

<u>POST-INFORMED CONSENT</u>

I ________________ ______________________________ agree to take part in research **"Topological relationships with children in the early years of geography teaching".**

I understood that I could take part, but that I could withdraw at any time without any problem. The researcher answered my questions and spoke to my carers. I have received a copy of this consent form and agree to take part in the research.

Itapejara D'Oeste, 2017.

____________________	____________________
Signature of the minor	Researcher's signature

ANNEX III

Informed Consent Form

"Topological relationships with primary school children in Geography teaching"

Dear Teachers:

We would like to count on your participation in the research **"Topological relationships with children in the early years of Geography teaching"**, to be carried out at **"Escola Municipal Nereu Ramos, located at 601 Rua Rui Barbosa, Centro, Escola Municipal Irmão Josafat Kmita, located at 365 Rua Canelinha, Bairro Industrial, Escola Municipal Prof. Pedro Viriato Parigot Souza, located at 2555 Rua Guarani, Bairro Guarani and Escola Municipal Escola Municipal Rural Valentim Biazussi, located at 2555 Rua Francisco Salviato Parigot Souza, Bairro Guarani. Pedro Viriato Parigot de Souza, located on the access road to Rua Guarani, 2555, Bairro Guarani and Escola Municipal Escola municipal do campo Valentim Biazussi, located on Avenida Francisco Salvi, 903, Barra Grande"**. The aim of the research is to **"analyse how topological relationships enable children aged 5 to 6 to perceive and understand geographical space. Your participation is very important. It will take the form of** interviews with questions aimed at teaching topological relationships - front/back/right/left, photos, transcriptions of speeches and descriptions of pedagogical actions.

Your participation is voluntary and you can ask to refuse or withdraw from participation at any time, without this entailing any burden or damage. We also clarify that the information will only be used for the purposes of this academic-scientific research and will be treated with the utmost secrecy and confidentiality, in order to preserve your identity. The names used will be fictitious.

This research meets and respects the rights set out in the Statute of the Child and Adolescent - ECA, Federal Law No. 8069 of 13 July 1990, which are: life, health, food, education, sport, leisure, professionalisation, culture, dignity, respect, freedom and family and community life. We also guarantee that Article **18 of the ECA** will be complied with: "It is everyone's duty to ensure the dignity of children and adolescents, keeping them safe from any inhuman, violent, terrifying, vexatious or **embarrassing** treatment. "

Should you have any questions or require further clarification, please contact: Dulcinéia Cristina Chiapetti, Rua Sete de Setembro/650 Barra Grande, telephone 999157337 or 35266061, email: dulce chiapetti@outlook.com. This form must be completed in two copies of equal content. **Itapejara D'Oeste,** 2017.

Dulcinea Cristina Chiapetti

ID: 5497232/6.

______________________________________ **(NAME OF REGENTING TEACHER)**,

having been duly informed about the research procedures, I agree to **voluntarily**

participate in the research described above.

Signature: ______________________________

Date: ___________________

Printed by Books on Demand GmbH, Norderstedt / Germany